INSECTS AND PLANTS
Parallel Evolution and Adaptations
(Second Edition)

by

PIERRE JOLIVET

SANDHILL CRANE PRESS, INC.
Gainesville, Florida, U.S.A.
1992

Acquisitions Editor: Ross H. Arnett, Jr.
In House editor: Renée Serowski

ISBN 1-877743-10-0

A FLORA & FAUNA HANDBOOK

LIBRARY OF CONGRESS CATALOGING IN PUBLICATION DATA:

Jolivet, Pierre, 1922-
    [Insectes et plantes. English]
    Insects and Plants: Parallel Evolution and Adaptations/by
    Pierre Jolivet. - 2nd ed.
    Translation of: Insectes et plantes.
    Includes bibliographical references and index.
    ISBN 1-877743-10-0 : $24.95
    1. Insect-plant relationships. 2. Insects - Evolution. 3. Plants-Evolution. 4.
Coevolution. I. Title. II. Series.
QL 468.7.J6512 1992
575-dc20   92-21646
                                                  CIP

MANUFACTURED IN THE UNITED STATES OF AMERICA

FOREWORD

Coevolution between insects and plants is a recent concept, widely discussed in the literature of the last decade, mostly in the United States of America. Coevolution may be defined as the synchrony of mutual adaptations between plants, usually, but not exclusively, green plants, and animals, especially insects. It is really a kind of action-reaction, the animals being the aggressors and the plants the defenders.

The term "phytoentomology" seems suitable to me as a term to properly express the contents of this book because this term summarizes the study of the facets involved in food selection by phytophagous insects and the reaction of their plant hosts. We have treated in this book, albeit rather superficially, many topics pertaining to agricultural pests and the agricultural side of this subject in general. This then, enables us to concentrate on purely biological problems such as symbiotic associations between insects and plants, pollination, carnivorous plants, and other topics mainly of biological rather than applied interest.

With the exception of several books on carnivorous plants, the first of which was written by Darwin, insect-fungus associations, pollination, and many symposia on the coevolution of insects and plants, there are surprisingly few books providing a comprehensive treatment of the subject of phytoentomology. I hope this modest contribution will help biology students and laypersons to better understand the mutual relationships between the two kingdoms of living organisms involved. At the same time, the applied entomologists and horticulturists will find here a discussion of the basic aspects of this science.

Naturalists interested in this field find it more difficult than those who work in laboratories where there is a considerable amount of experimental data available. However, field naturalists are able to directly observe from the open book of nature, which, particularly in the tropics, provides rich and seemingly unlimited data. Many of the observations included here come from field work in both the New and Old World tropics, chiefly in the latter, during my more than 30 years of study.

The French edition of this work appeared in 1983, and has since gone out of print. Meanwhile the first printing of the English edition of the work was first published in 1986, and it too soon sold out. Because of the continued demand for the work, this second printing, with a few corrections and additions is offered herewith, lack of time preventing a complete revision of the work, which remains to be done in the future if this book continues to serve its purpose.

-Brazil, April 6, 1992

PREFACE

Pierre Jolivet's book, "Insects and Plants," is on one of the most attractive subjects a biologist can offer to an educated public. The erudition found in this book is softened by the simplicity and clarity of his text, which makes the book readable by all. In the world of living things, cooperative correlations between two or more species are frequently observed. More than half the species of insects are the oppressors of the plants on which they feed. Less commonly, the reverse is true. Even more, certain plants are dependent on insects for survival. Information on the latter surpasses the former in interest and complexity. The herbivorous insect does not eat just any plant that it comes across. Even the migratory locust, despite its well known insatiable appetite, has its preferred host plants, and avoids other species even to the point of starvation.

Generally insects will feed on numerous species of plants; sometimes they are restricted to a single family, a single genus, or even a unique species. These insects are monophytophagous. It is now known that plants emit odors which attract certain species of insects. The classical example of this is the attraction of the cabbage butterfly to mustard oil of the cruciferous plants. Another example is the terpenes present in conifers. These substances attract scolytid beetle pests of pines, firs, spruce, and other trees.

It is not to discover the diet that best suits the species, but to find the factors which have modified the genetic code of the species for this specialized way of life that is difficult to determine.

Odor from food plants is a significant stimulant for insects. Molecules of substances emitted from the plant, as photons emitted from a light source, go out in all directions. The few molecules which reach an insect's antenna are sufficient to orient and direct the insect to its favorite plant host. There it may feed, or, if a female, it may lay eggs.

The study of these phenomena is a study of evolution. All of the facts have not yet been found. For example, how and when did transformations from a general to a specialized diet take place? In the example above, a minute quantity of molecules (emitted from the specific plant and difficult to detect) initiates the movement of the insect toward the stimulus source.

Carnivorous plants are amazing because of their specialized method of getting food. They have acquired in the past the ability to feed like animals. Pierre Jolivet describes with precision their specialized feeding mechanisms. These plants, even though they have this strange ability, retain their autotrophic characteristics, including the chlorophyll needed for photosynthesis, and they probably could survive for a time without any animal food as their only nitrogen source. The movement of leaves and other plant organs, anatomical peculiarities adapting them for predation, the digestive enzymes, and the permeability of cell membranes to digested organic products, equips the approximately 500 species of carnivorous plants known today for this strange way of life. Their movements are coordinated to the point of operating efficiently as passive predators. The evolution of these plants, belonging to a few isolated families, has come about through preexisting genes, although the necessity of this revolutionary change for the survival of the species does not seem particularly evident. These details may be found in this book.

The most complex evolutionary problem discussed in the book concerns the coevolved interrelationships between plant and insect. Many plants must be fertilized during the visits of pollen collectors or nectar feeders. When a flower is pollinated only by the intervention of an insectivorous insect, one or the other, or both, must be greatly modified for this purpose. The book describes with precision the anatomical features and the chemical attractants utilized by the plants in this relationship with insects (see chapter 12).

It is difficult to explain these unique lines of development; it is much more difficult to follow and interpret the coevolution taking place in two distinct, widely separated species. One is modified to compliment the other. Unsuspected properties are put to work to achieve this harmony.

Two views on coevolution are expressed by biologists. For some, the plant and the insect evolved separately. It is just a coincidence that this happened and has nothing to do with a special evolutionary process. It happened to be more or less useful to the plant. In fact, they say, anthropomorphism is the inventor of this so-called evolution. On the other side, the view is that the

insect and plant evolved together, each receiving aleatory mutations, but keeping only those which gave the plant or the animal
an advantage. This is the application of simple Darwinian sele-
cion. Darwin was, during his life, perplexed by reports of this
kind of insect-plant relationship. In fact, he devoted an entire
book to one aspect of this, a book on carnivorous plants.

But of what value was Darwin's and his followers' explanation? The postulation of so many required reciprocal genes and
mutations, and that in two opposing lines of development, makes
the probability of such events extremely remote almost to the
point of being nonexistent. Other conditions needed to permit this
double plant-insect evolution further weakens the aleatory thesis.

Even though more and more data is being added each day, for
the most part, the present situation requires a biologist to have
the courage to say "I don't know." Acknowledging this fact does
not lower his esteem, but rather, honors him. Pierre Jolivet's
totally objective book opens several rich and vast fields to explo-
ration. It stimulates reflection. The biologist must not limit his
studies to a few cultivated or cultured species. He cannot have a
true idea of the complexity of biological phenomena without
considering animals and plants together in their natural habitats
and observing their mutual interactions. The living world is
brightened by its diversity. The questions suggested to the observer and thinker are of a quasi-infinite number. Only a very
small part of them have held the attention of scientists.

Reckless and arrogant is the one who pretends, in the pre-
carious and limited state of our knowledge, to explain the fundamental mechanisms of life and the avatars of its forms.

Pierre Jolivet's book, even with its richness of facts, clearly
shows our ignorance. It is a book of truth. What better can be said
of a scientific book?

-Pierre-P. Grassé
Member of the French Academy of Sciences,
Professor of Zoology

CONTENTS

"And what does it live on?"
"Weak tea with cream in it."
A new difficulty came into Alice's head.
Supposing it couldn't find any?" she suggested.
"Then it would die, of course."
"But that must happen very often,"
Alice remarked thoughtfully.
"It always happens," said the Gnat.
-Lewis Carroll

INTRODUCTION

If the diet of Alice's companion was so selective, the diets of many phytophagous animals, the herbivores, are no less strict. It is evident all degrees of food selection exist, from the insect which eats only one species of plant, to the cow which eats almost all green herbs (with some notable exceptions, however), and the slug which also eats most anything. Food selection evolved among all animals, not just among these phytophagous creatures.

Mosquitoes, chiefly anopheline, for instance, are anthrophilic or zoophilic, since they prefer animal, even human, blood. Not in every case are the animal hosts bitten at random. Several *Anopheles* species will bite here a pig, there a cow, a goat, or buffalo, but they are still selective. We have often seen *Anopheles minimus* in the Philippines, a stream breeding species, preferably biting the local variety of buffalo, the Carabao (*Bubalus bubalis*), while the European domestic cow is consistently rejected. This phenomenon is comparable to the peculiar behavior of some phytophagous insects that reject imported plants, a phenomenon to be explained later, called "xenophobia."

Similar food selection patterns exist among all living things. It is exemplified in some species of vertebrates (for example, insectivores, and Galapagos finches) narrowly adapted to various specialized ecological food niches. At the same time, few animals are really omnivorous. Even our own species, *Homo sapiens*, with our limited digestive apparatus (*e.g.*, impossible to digest cellulose) and gustatory preferences, seems to have developed various "phenotypes" conditioned to various cuisines: macaroni (Italy), pimento and curry (India, Sri Lanka), rice (Far East), teff (*Eragrostis teff* [Gramineae], also called Love Grass, Ethopia), kimchi (Korea), and there are others.

Culinary preferences and religious taboos also contribute to the restrictive food habits of the human species; for example, forbidding horse meat (in Western Europe), mutton (in Holland); pork (in Israel, and the Arab countries); beef (in India); a great variety of animals, including pig, rabbit, and duck (in Ethiopia), and all meat in several Buddhist and Hindu religious sects. The most polyphagous human beings are surely the Chinese who have succeeded in extending to the maximum the alimentary spectrum, animal and vegetable alike. Most of these taboo barriers are purely subjective. Some, however, evolved as health protection, for example, forbidding pork because of trichinosis worm parasites. These gustative "phenotypes" however, are conditioned from infancy and are very difficult to modify at a later age.

If we think about it, food available to humans depends on geography, climate, and season. In most of the world, cheap foodstuffs, such as rice, are the basis of the diet. Other diets consist mainly of wheat and potatoes as in Europe, or corn in South Africa, east of Sudan, and in Mexico; cassava in the Congo and in southern India; rice in southern India; teff in Ethiopia; dates in southern Iran, and sweet potato in New Guinea. Many of these are not even indigenous to the countries cited. Before the diversification of agriculture, the main sources of protein differed greatly among the countries: beef in Europe, sheep in the Middle East and other Arab countries, pork in the Far East, fish in Japan, insects in Thailand, and lizards, snakes, and frogs among Australian aborigenes. Anthropophagy as a source of protein is, happily, becoming increasingly rare, and recently has been recorded only among some isolated tribes in New Guinea. This is not at all a food specialization, but a supplementary contribution to their normal diet of pork. Real food specializations are very difficult to detect among those not at all conditioned by religious taboos, ways of life, money, climate, and other factors. Even if genetic and physiological factors determine food selection, at least at the individual or family level, new tastes have been created as well (cola soft drinks, for example). Adaptation to these foods is real, but biologically it is no more than acquiring a new habit.

The subject of food selection is discussed in the following pages, concentrating on arthropods, particularily herbivorous or phytophagous insects. The subject is so vast that only some aspects will be treated in detail; others will be mentioned only in passing. This is followed by a review of the whole gamut of insect-plant relationships, including such subjects as myrmecophily, carnivorous plants, pitcher plants, symbiosis, and pollination. References to these various subjects are given in the bibliography, including several general as well as specialized publications. However, no attempt has been made to list all of the recent publications. Those interested in further references should consult the works of Ahmad (1938), Balachowsky (1951), Brues (1946), Futuyma and Slatkin (1938), Painter (1951), Hering (1951), and Dethier (1947). Any one of these short chapters could be expanded into a whole book. The present book serves as a synopsis, and that is the only aim of the book, except, of course, we wish to reach beginning students of this subject. Sometimes comparisons between food selection patterns of phytophagous plant parasites, such fungi as Uredinea, and those of insects are made, as well as with those among such pests as spiders, molluscs, and nematodes. It is obviously impossible, however, in a book dealing with insects and other arthropods to give details about the diet of other phytophagous animals and parasitic fungi.

Agronomical aspects of plant-insect interrelationships are only superficially treated; there are enough books in various languages on this subject, as well as on the physiology and chemistry of food selection of insects.

This book was written during many trips in foreign countries, and, as mentioned previously, it has been enriched by personal observations.

I apologize for any imperfections and inevitable mistakes which may have been inadvertently introduced into the text.

So, naturalists observe, a flea
Hath smaller fleas than on him prey
And these have smaller still to bite them.
-Swift

CHAPTER 1

Diets of Living Things

The diets of living organisms are grouped by most authors in the following categories, but as we shall show in the following chapters, organisms may not be restricted or limited to any one of these.

Holozoic organisms - Animals that feed on organic substances, engulf solid food particles, break these down to simple compounds during digestion, and then absorb, assimilate, and oxidize them for growth and energy. To do this, these organisms must prey on other animals, or they must feed on plants. With rare exceptions, all animals are holozoic, although the term may be restricted to free living animals (see also parasitic animals which follow).

Autotrophic (=holophytic) **organisms** - Green plants, certain bacteria, and a few protozoa use the compounds carbon dioxide, water, and certain minerals, through the action of sunlight, to synthesize carbohydrates, which is, in turn, used to manufacture fats and proteins. These are oxidized for the release of energy, or synthesized, through growth, into living tissue. In other words, these organisms obtain food through the process of photosynthesis.

Saprophytic organisms -Fungi, bacteria, and some plants that live on other plant and animal tissues, living or dead, utilizing them either directly, or after some external digestion (reduction), by enzymes. Organisms devoid of chlorophyll, such as fungi, most bacteria, and certain flowering plants, are well known saprophytes. In this group one can, in a sense, place carnivorous plants even though they are almost entirely autotrophic. Several tropical and a few temperate flowering plants entirely lack chlorophyll, making them dependent for nourishment, as are fungi, on decaying vegetable matter. These plants vary in color from white to brown. The species fall in various plant families such as Triuridaceae, Gentianaceae, Orchidaceae, and Polygalaceae (e.g., *Burmanrua* spp.).

Saprozoic organisms - Animals that survive only on dead plant or animal matter are termed saprozoic. This includes saprophagous, necrophagous, and coprophagous groups, such as numerous protozoa, insects, and abyssal fishes. Coprophagous animals, those that feed on feces, are saprozoic because of their animal food source. The phylum Pogonophora, which is completely devoid of a digestive tract, is both saprozoic and saprophytic because they absorb nutrients from the sea bottom directly through their integument. These mixed diets occur in certain other marine, fresh water, and terrestrial species.

Parasitic organisms - Living organisms, animal or plant, that get their food partially (hemiparasitic) or totally (holoparasitic) from another individuals without necessarily killing it, are termed parasites. These absorb blood (serum or red cells), sap, digested food, or the tissues of its living animal or host plant. Parasites may be internal (endoparasites) or external (exoparasites). A "good" parasite, well adapted to its host, must never cause any fatal disease. If it kills its host, it, in effect, commits suicide.

Symbiotic organisms - Organisms associated by reciprocal processes, which is, in a sense, only a peculiar type of parasitism. We can consider this as being "the utilization of the parasite by its host," *i.e.*, it is a durable association between two organisms (plants, animals, or plant and animal), usually with mutual benefit.

Diets

Animals feed on complex substances, either of animal or plant origin, *i.e.*, substances already manufactured by plants from simple material. Animal diets may be classified in various ways. Animals feeding only on plants are herbivorous or phytophagous; those feeding only on meat are carnivorous; those which live, as do most humans, on a mixed diet of animal and vegetable materials are omnivorous; and those which feed on excreta or dead matter, generally by absorbing it via the digestive tract, are saprophagous.

When animals attack and eat other animals they are called predators. Parasites (either endo- or ectoparasites) are "predators" which are smaller than their host. Many do not necessarily kill their host, at least not at the beginning, but others eventually cause the death of the host. Parasites are highly specialized morphologically for their mode of living. Sometimes they are themselves parasitized by other parasites, termed hyperparasites, which are sometimes attacked by still other parasites, the hyper-hyperparasites. This third degree parasitism is best shown in some parasitic Hymenoptera. Recall Swift's famous verse at the head of this chapter. This phenomenon may be written in different words, but it always has the same meaning.

Whether the animal is parasitic or free-living, carnivorous or herbivorous, plants, the primary producers, form the base of the food chain of all living things. Food chains are rarely more than four or five steps from the plants to predators and their prey (fig. 1). In simple terms, a plant is eaten by a herbivore which is, in turn, eaten by a carnivore. The carnivore is eaten by another carnivore, and so on. Classic examples of such chains are, for instance, the relationship between plankton-crustaceans-fishes-man or the relationship between grass-gazelle-lion.

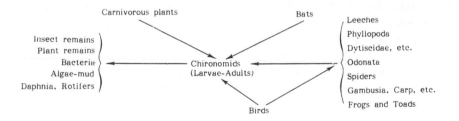

FIG. 1. An example of a food chain with a pantophagous species: chironomid larvae (after Toureng, modified).

Humans, according to Burnett (1959), are dependent for life giving food on "six inches of soil." It is indeed evident that our species feeds on plants and herbivorous animals which feed on plants. That is also true of all living things on our planet. Many carnivorous species also depend directly on plants for at least a part of their food, such as their need for Vitamin C. Others find in their diet of herbivorous animals all their nutritional needs.

Food chains show a progressive diminution of the number of individuals involved and an increase in the size of the animals at the end of the chain. In other words, the number of predators becomes smaller and smaller as the size of the animals at each stage increases (fig. 2a).

In parasite food chains, however, the pyramid is completely inverted. Starting with a single organism, the host, the parasites become progressively smaller and smaller in size, and more and more numerous at each level (fig. 2b).

Niche

The ecological term "niche" means literally the way by which an animal obtains its food, *i.e.*, an "ecological niche" can be compared to a profession or a trade in human society. A niche is characterized by the kind of food used, which in turn is determined to a great extent by the size and adaptation of the animal itself.

Let us take, for instance, the description by Sankey (1980) of life in an European woods. Here, the predatory niche is occupied by foxes, badgers, and weasels among the mammals, and by falcons and owls (day and night predators respectively), among the birds. Other carnivorous niches may be found: shrews, which feed on invertebrates; birds which eat ladybird beetles (which in turn eat aphids); frogs which eat insects; and carabid beetles which eat slugs. These niches are distinguishable from each other by size and the various food requirements of these predators. Similar niches are found among herbivorous insects and mammals.

Nutrition

As we have seen at the beginning of this chapter, plant nutrition is not significantly different from animal nutrition. In both cases, there is food gathering and digestion, its subsequent use as an energy source, or its assimilation as living protoplasm. There exist, however, between the two kingdoms, fundamental differences in the method of obtaining food, in digestion, and in transportation of the food to the cells. Therefore, it is necessary to outline the main divisions of food acquisition and nutritional systems among the plants and animals.

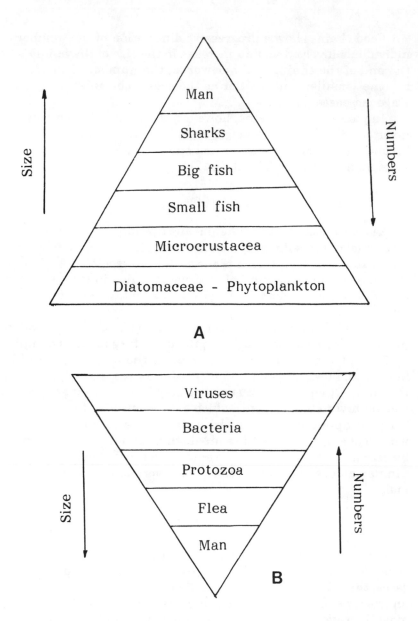

Fig. 2. Examples of food chains: A- ordinary food chain, normal pyramid; B- parasite food chain, inverted pyramid.

I. Nutrition among plants.

Many plants are autotrophic, *i.e.*, they are able to produce their own food through photosynthesis. However, other plants are heterotrophic, *i.e.*, as in animals, they obtain their food from sources outside their own body. We can divide the heterotrophic plants into saprophytic, parasitic, and insectivorous, and the autotrophic plants into chemiotrophic and phototrophic.

Heterotropic plants

Saprophytic plants – Many of the lower plants, such as fungi and bacteria (if we consider them to belong to the plant kingdom), and many protozoa, which are devoid of chlorophyll, feed by saprophytism, *i.e.*, from organic substances dissolved in water or from dead organic matter which they digest outside their own body. Many saprophytic organisms, struggling for vital space, secrete antibiotics to exclude each other. We have learned to use some of these antibiotics, those secreted by certain fungi, as medicine.

One group of marine animals, the Pogonophora (see p. 5), are the only metazoans known to feed by absorbing food externally. They totally lack an intestinal tract and depend entirely on liquid nutrients present in the sea.

Parasitic plants – Parasitic plants obtain their food directly from the tissues of other plants or animals. As we have seen, these plants can be divided into holoparasitic, *i.e.*, totally parasitic and completely devoid of chlorophyll (flowering plants like *Rafflesia*, dodder, the tropical Lauraceae, *Cassytha*, fungi parasitic on plants and animals, several parasitic algae or algaellike organisms, all the bacterial parasites), and hemiparasitic or semiparasitic with chlorophyll (as the mistletoe, *Viscum album*, and other Loranthaceae, as *Orobanche* sp., etc.). Parasitic underground orchids (in Australia) are devoid of chlorophyll, but they are also partly symbiotic with fungi.

Insectivorous plants – These plants will be more fully discussed later since they are a good example of the relationship between insects and plants. It is well known that many insects feed on plants, but in these insectivorous plants a completely reverse phenomenon occurs. Insects are digested and assimilated by plants specifically adapted to this purpose. The best known of the carnivorous plants are species of *Drosera, Dionaea, Nepenthes, Sarracenia,* and *Darlingtonia.*

Several diastases, enzymes that help in digesting captured insects, are utilized by these plants. The digestive process is often helped by bacterial fermentation in pitcher plants. The nitrogen low soil of these acid peat bogs is made up in this manner. All the carnivorous higher plants retain their chlorophyll and their autotrophic properties. Nitrogen derived from the digested insects apparently is the slight compensation they receive from this digestion. They can be considered as "hemiparasitic" on insects. Moreover, around 20 species of fungi devoid of chlorophyll also have trapping mechanisms. They trap mostly soil nematodes.

Autotrophic plants

Chemiotrophic plants-Many bacteria and several algae or algaelike organisms (Cyanophyceae, etc.) are able to obtain the energy they need for the synthesis of their own protoplasm (Chemiosynthesis) by oxidizing several elements or inorganic compounds such as hydrogen, iron, and ammonia, and by combining these with carbonic acid and water.

Phototrophic plants-Phototrophic plants (the higher green plants, Rhodothiobacteria, and the chlorophyllic protozoa, such as some *Euglena* spp.) are able to use sunlight as energy for synthesizing their own organic matter from carbonic acid, water, and mineral salts. Chlorophyll, and various other pigments among brown, red, and blue algae, are able to absorb light energy from the sun to induce photosynthesis.

II. Nutrition among animals

With the exception of some chlorophyllic protozoa such as phytoflagellates and several amoeboids, all animals are heterotrophic, *i.e.* they obtain their food already synthesized in the form of carbohydrates, lipids, and proteins. Unicellular autotrophic organisms are really intermediate between green algae (Chlorophyceae) and the free-living or parasitic protozoa devoid of chlorophyll. Normally, all higher animals feed on complex organic matter.

Due to their extreme complexity, it seems one can classify only with difficulty, the various diets of the higher animals. However, the following simplified classification is proposed.

Carnivores

Predators- All predators are flesh eaters. These may be further divided, according to the nature of their prey, into insectivorous, piscivorous, *etc.* Among the insects, at least, it may be that the carnivorous food habit is derived from the phytophagous diet via saprophagous habits, but there is no direct evidence of this. Carnivorous animals, such as humans, lions, and most carabid beetles, hunt their prey. People are only partially predators. Many carnivorous animals, chiefly insects, are narrowly adapted to a specific kind of prey.

We know that the bigger the carnivorous animal, the wider its hunting territory. Because prey is not always available when needed, carnivorous animals gorge themselves whenever they can at irregular intervals in contrast to the herbivorous species which are able to eat on a regular schedule. Humans, during prehistoric times, were obliged to eat when they could, when they captured prey. Even today, in isolated areas, some migrating Inuits fast for long periods of time between enormous meals. Both herbivorous and carnivorous feeding habits have produced, among the big animals, different adaptations of their digestive habits and their intestinal morphology.

Hematophages-These are blood suckers, such as mosquitoes, fleas, some Hemiptera, and other animals which pump out the blood of vertebrates or the haemolymph of insects. Among the mammals, one American bat, the vampire bat (*Desmodus* sp.), does much the same, lapping the blood seeping from wounds which they make on cattle, horses, and even humans. The majority of bats, however, are insectivorous, frugivorous, or piscivorous. Another example of blood sucking animals are the leeches, which are generally aquatic, but some in Southeast Asia are terrestrial. The blood feeding insects, including *Cimex* spp., Reduviidae, and fleas, are also irregular eaters. Each time they do eat, their abdomens inflate considerably.

Parasites-It is useful to distinguish between the externally (ectoparasitic) and internally (endoparasitic) parasitic animals. Parasites must be viewed as carnivorous since they partially eat their host's tissues. Leeches, which live chiefly as ectoparasites of vertebrates, as we have seen previously, are also hematophagous as well. Parasites, particularly Hymenoptera and Diptera, whose larvae completely consume their host, but which are free-living in the adult stage, are called "parasitoids." Some other parasites, such as the Nematomorpha, the Gordioidea worms, also parasitic only in the larval stage and free-living as adults, are termed "protelian parasites." Many other worms (Trematoda and fleas) are free-living as larvae but parasitic in the adult stage.

Filter feeders (on animal plankton or macroplankton)- Many aquatic animals, marine, or freshwater have special mouthparts designed to filter out microscopic or very small plants and animals as their food. Among them are passive feeders such as sea anemones and corals which strain out all organic matter while remaining in one place. Others, such as certain fishes, whales, and most crustaceans, however, are active feeders, moving about to find their food.

Herbivores or phytophages

The easiest way for an animal to obtain food is to eat green plants; it is also the most common way of feeding. Even in the Arctic tundra, reindeer eat lichens and in the desert and semi-deserts, there are many species of plants which are the food of both wild and domestic animals. Even in the most severe of deserts, there may be edible lichens, (particularly *Lecanora* sp.) blown there by the wind, attached to rocks.

To separate the herbivorous animals into distinct categories is very difficult. The following classification is purely morphological and neither botanical nor chemical. Later in chapter 2 a more scientific classification is proposed based on plant species and kinds of chemicals.

Filter feeders (on phytoplankton=microplankton)- Many small aquatic animals feed mainly on Diatomaceae and various other microscopic algae. Many mosquito larvae, for instance, feed chiefly on bacteria and plant microplankton (Diatomaceae, Dinoflagellates).

Lichen and algae eaters- Numerous aquatic invertebrates feed on algae. Lichenivorous animals are found among the insects and the vertebrates. Humans occasionally eat lichens, and green, red, and blue algae.

Fungivores or mycetophages- Fungus eaters can be external feeders such as slugs, or internal feeders such as Mycetophilidae larvae and several groups of beetles. Other insects feed on the underground rhizomes of fungi.

Moss eaters- A few insects feed exclusively on Bryophytes. Certain Diptera eat Hepatica, the liverworts, and certain Tipulidae, the crane flies, and certain chrysomelid beetles, as well as some Lepidoptera larvae, are exclusively moss feeders.

Microphages- Insects known to eat minute living plants such as yeasts and bacteria, or plants stages such as spores, and pollen (see also floricolous animals), are microphagous. They differ only from plankton eaters in that they are terrestrial. Most Thysanoptera and many Coleoptera species and others can be classified as microphagous and floricolous insects. Many other insects considered to be saprophagous are really microphagous feeders.

Leaf eaters- Here one can distinguish between the species which eat leaves (most animals) and those which mine them, digging galleries between the palisade and lacunous tissues or within one of them (some Diptera, Coleoptera, and Lepidoptera).

Sap suckers- All of these, including aphids and other Homoptera, as well as Hemiptera, are insects with piercing-sucking mouthparts. Sap suckers have evolved in several instances into blood suckers (*e.g.*, some Reduviidae).

Frondicolous animals (bud and shoot eaters)- Such specializations are not only found among insects, but also among more evolved animals. Gorillas, for instance, in the Kivu mountains of the Congo feed mostly (but not exclusively) on bamboo shoots, as do pandas in western China. The Australian koala prefers the young shoots of certain species of eucalyptus.

Floricolous animals- These are the flower eaters. Some specialize as pollen eaters, such as bees and certain beetles: Cetonidae, Orsodacninae, some Galerucinae, some Meloidae, all Oedemeridae, and others. Certain Thysanoptera or thrips belong to this category also, as well as several groups of Diptera and Hymenoptera.

Miners (in shoots and stems of herbaceous plants)- These insects dig galleries only within non-woody flowering plants.

Xylophagous animals (wood eaters)- These animals mine in the wood of trees or under the bark of dead or dying trees (Scolytidae, Bostrichidae). Some are not really xylophagous but rather, mycetophagous. Xylophagous insects such as primitive termites, cockroaches, and beetles raise an interesting physiological problem: there must be yeast, bacterial, or protozoan symbionts present in their digestive tracts, often in specially differentiated sites. These digest cellulose for their hosts. The sea mollusk, or shipworm, *Teredo navalis*, although feeding mostly on small algae, also pierces and eats wood, which seems to be for them, an ideal food. The shipworm was thought not to possess any symbionts, but actually it has symbiotic bacteria, found recently, which help in the digestion of wood. Among the species of shipworms examined, all were found to possess these bacteria. Hematophagous insects also have symbionts to help them digest blood.

Frugivorous animals- These are the fruit eaters, ranging from birds to Coleoptera adults and larvae, Lepidoptera, Diptera, and others. We can distinguish the external feeders from the internal ones. Numerous exotic bats feed only on tropical fruits, usually figs.

Spermophages or Clethrophages- These are seed eaters (some Curculionidae, Bruchidae, some Tenebrionidae, and others), which also attack stored foods, grains, and flour. Some birds have an exclusively spermophagous diet. They can cause severe damage to rice paddies, particularly in Africa. Some ants are known to feed on the seeds of grasses which they accumulate and store in their nest galleries. Often Curculionidae and Bruchidae that are spermophages are monophagous, feeding only on a single species of seed.

Honey gatherers and nectar feeders- These species have either a sucking proboscis, as Lepidoptera, or certain Coleoptera, or they lap nectar as do the bees. Certain birds equipped with a specially adapted tubular tongue, such as hummingbirds of the New World and the sun birds of the Old World, feed mainly on nectar. They may also eat insects and spiders that are in the corollas. Nectar feeders are remarkable in always having a special, local species adapted to the flowers of trees, bushes, and plants on which they depend for food. Some bats are also very fond of nectar. Well known coadaptations exist in various species in the tropics, for instance between day-flying birds and species of *Rhodendron* in New Guinea.

Radicicolous animals (root eaters)- Some of these feed within roots, *e. g.*, larvae of Eumolpinae beetles (Chrysomelidae). Others feed externally on the root, *e. g.*, the many subterranean insects that dig galleries in the soil (*e.g.*, *Gryllotalpa* spp., tropical ants, root feeding endogeous aphids, and some Alticinae adults and larvae).

Gallicolous animals– These are animals which produce and live within galls, outgrowths on the stems, leaves, flowers, or roots of plants. Galls, or cecidia, are in reality benign tumors which form in plants around the developing parasites. They are classified as zoocecidia, mycocecidia, or bacteriocecidia, according to the nature of the parasites, *i.e.*, animals (nematodes, Acari, insects), fungi, or bacteria. The subject of gall and gall making will be discussed in chapter 9, but it should be pointed out here that the roots, leaves, stems, and flowers of many green plants bear specialized galls for each species inducing this abnormal growth.

We can note here the strange mode of feeding of honey ants (Camponotinae and Dolichoderinae), which collect chiefly honey-like sweet secretions from galls on certain bushes made by a chalcid wasp. This honey is stored within the abdomen of special worker ants, known as repletes, which hang from the ceiling of the nest. Some of this honey also is collected by the ants from the excreted honeydew of aphid and scale insects. It is very difficult to decide whether to classify those ants as sap suckers or co-prophages.

Saprophages

These feed on dead organic plant or animal matter.

Coprophages– Some insects such as many scarab beetles, feed only on the excreta of vertebrates and probably also on the microflora of the excreta. They are more or less selective in the origin, age, and consistency of the excreta. Jeannel (1946) said that those coprophagous insects which follow ruminants or Equidae have essentially a vegetable diet. This does not seem completely true if we consider that those insects also digest the intestinal symbiotic ciliate protozoa, which, therefore, constitutes a meat diet. It is also evident that some beetles are very special-ized in their choice of excreta, being restricted to the dung of one or a few species. Others are less selective and even may be par-tially necrophagous (*Trox* spp.). Among the saprophages are populations of guanobites feeding on excreta in bat caves. The woodlice (Crustacea) feed on excreta deposited in wall crevices.

Detriticolous– These are animals that feed on dead or decaying organic matter of animal or vegetable origin. Among the insects we note Apterygota, numerous Diptera, Coleoptera, and others.

Many insects, such as *Drosophila* spp., said to be saprophagous, feed mostly on fungi and bacteria which multiply on
detritus and fruits. Marine worms, the Polychaeta, sea cucumbers
(Holothuria), and others, are detritus feeders. They swallow huge
quantities of sand and mud, and as with the earthworms, extract
all the nutritive substances from the mixture, rejecting the rest.

The distinction between herbivorous and saprophagous species among plankton-filtering animals is rather difficult. The size
of the food particles more than their nature decides the final
selection of food by the animals (Rotiferae in fresh water, for
instance). Among detritus feeders in fresh water, we find in the
mud mollusks (species of *Unio, Anodonta*), Diptera larvae (*Chiro-
nomus* spp.), annelids (*Tubifex* spp.), and others.

Those abyssal fishes that are not carnivorous are detritivor-
ous. They feed on the bottom mud, rich in animal or vegetal
organic matter, that drops down from the surface.

Necrophages- These are the animals that feed on dead
animals. The hyena, jackal, and vultures almost exclusively feed
on dead carcasses, even if they are able to kill their own prey. On
the other hand, purely carnivorous animals, like lions, rarely
(particularly old animals which no longer hunt) feed on dead or
decaying animals, as for example, those that I observed once in
Albert National Park in Zaire feeding on dead hippopotami.

A remarkable adaptation among necrophagous insects has
developed in the Holarctic beetle genus *Nicrophorus* (Silphidae).
Necrophagous species are relatively specialized. Among the insects, species of *Necrophorus* are derived from carnivorous or
coprophagous species which became specialized on a diet of vertebrate or insect carcasses. We find necrophagous species among
certain families of Coleoptera and Diptera. Some are only attracted by rotting flesh, others by drying or mummified tissues and
bones, or by fats, hair, horns, or feathers. But others, for instance,
a big species of *Trox* from Africa, are known to feed indiscriminately on human excreta and on mummified carcasses of large
vertebrates. It is not certain if different species are involved.
Many seem to especially prefer skin and hair. The distinction
between coprophages and necrophages is rather subtle and both
diets are rather similar. Due to their various specializations,
which are shown by different tastes and attractions, necrophagous insects succeed each other in teams, according to
nature and the degree of putrefaction of the corpse. Specialists
can rather accurately determine the day of death based on the

species of insects present.

Omnivores or pantophages
Omnivorous animals- These species feed on plants and animals, dead or alive. Human beings are one example. Some of them are also partly saprophagous. Among the insects, pantophagous species, those which have a varied diet, but do not eat everything, are not to be confused with polyphagous species.

Alternative diets
In some cases, some animals are herbivorous as larvae and carnivorous as adults, or *vice versa*. Good examples of this are hydrophilid beetles, and predatory Hymenoptera that are carnivorous in the larval stage and exclusively herbivorous in the adult stage. The converse exists with the tadpoles of several Amphibia which are herbivorous, while the adults feed on insects, crustacea, worms, small mollusks, *etc.* The digestive tracts of the herbivores are always longer than those of the carnivores. The tadpoles of the Anura (frogs and toads), have a very long intestine which is rolled into a spiral. It is adapted for the digestion of purely vegetable food. When metamorphosis takes place, the intestine shortens considerably as the diet changes. Butterfly caterpillars are herbivorous; the adult is usually nectariphagous, hence, both diets are phytophagous.

In allotrophy, a subject which will be discussed later as it occurs among the phytophagous insects, carnivores can become herbivores or frugivores (*e.g.*, *Carabus* spp.) or exclusively leaf feeders. For example, some leaf beetles may show cannibalism by eating their own eggs and larvae. Diet may be modified under the pressure of necessity or climatic factors. For instance, during delivery, an otherwise herbivorous female rabbit may eat her own offspring.

The change from a meat to a plant diet, and *vice versa*, is often observed among insects, mostly among the facultative saprophages and some phytophages. Sometimes the food habits may be different for the male and the female of a given insect species. Such is the case of the species of *Culex, Aedes,* and *Anopheles* mosquitoes, where the male sucks nectar of flowers and the female sucks both nectar and vertebrate blood during the greatest part of her life. The first meal of an *Anopheles* female is a nectar meal, but the filling of the stomach with blood is necessary for the completion of the gonotrophic cycle.

However, it is true that *Culex* spp., according to the race involved (races of *Culex pipiens* among them) show some isolated cases of autogeny, *i.e.*, the development of eggs without a previous blood meal. The difference between the two races of *Culex pipiens* is as follows: one can lay eggs without a previous blood meal, and the other can be maintained during at least one generation on fruit juice alone. This ability depends on the richness of the proteic food received during the larval stage. No *Anopheles* species has, under experimental conditions, laid viable eggs without a previous blood meal. Despite that, recent mosquito experiments have again raised this question. The literature has a few reports of some *Anopheles* females sucking the sap of banana trees and even of blood sucking males, but these appear to be exceptions. Normally, the male mosquito is totally unable to pierce animal skin because of the atrophy of its mouthparts. We will see later that there are even blood suckers among certain groups of Diptera, thought to be otherwise totally phytophagous or aphagous (see below), as for example, some Chironomidae.

Aphagy, lack of feeding, is known among many insects. Adults of Ephemeroptera, some moths (Saturnidae), some Diptera (Chironomidae, several species and genera) and many other adults do not feed. They are unable to feed usually because they have no functional mouthparts, and they have also lost their digestive tract. Food accumulated during the larval life is sufficient to keep the adult alive during the short time spent mating and egg laying.

"Lucullus cenat apud Lucullum"
- Plutarch

CHAPTER 2

Food Selection Among Phytophagous Insects

Without having the refinement of Lucullus, quoted above ("Lucellus eats at home") the Roman gourmet, phytophagous (or herbivorous) insects are, however, very restricted when selecting their food. For instance, we have observed personally, in the mountains of New Guinea, caterpillars of the beautiful *Papilio ulysses*, the "morpho" of the Australian tropical region, dying on citrus leaves because the normal host plant in the lowlands is one species of *Evodia*, another member of the Rutaceae. This is an extreme case. Strict monophagy is rare at the species level rather than at the generic or family level.

We have seen previously that phytophagy among insects is rather irregularly distributed among the various orders. It also exists among some other arthropods, including several arachnids, and some myriapods. The Crustacea are polyphagous, some carnivores or detritus feeders, and although some are terrestrial, only the pagurides, (for example, *Birgus latro* and *Cenobites* spp.) are almost entirely phytophagous. The coconut crab feeds exclusively on coconuts in the South Pacific and can be raised in captivity on boiled rice. This is a common practice in the Philippines and other areas.

Phytophagy occurs in various forms: as external phytophagy (feeding on stems, leaves, flowers, fruits, and roots), or internally as in the case of gallicolous species, leaf miners, wood borers, and fungus feeders. Detriticolous and non-carnivorous myrmecophilic species of Clytrinae (Chrysomelidae, Coleoptera) feed mostly on detritus, algae, and fungi. We can find phytophagous species in a wide variety of habitats, even in caves, for example. Several species of some Homoptera, usually stem and leaf feeders, feed on roots and other variations are not unusual. Many insects usually regarded as phytophagous are actually mycetophagous and feed on molds. Bruchidae (Coleoptera) and other seed feeding insects, might be regarded as phytophagous in the broadest sense, because differences between miner, gallicolous, and spermophagous species are rather difficult to delineate, especially when one must deal with big beetles, for example, the Sagrinae (Chrysomelidae, Coleoptera) that feed inside of leguminose stems.

The rules of phytophagy that Maulik once tried to codify, are applicable, not only to insects, but also to other selective phytophagous organisms such as Uredinea, the rusts, other parasitic fungi, and Acari (mites). This classification is certainly imperfect, but what classification accounts for all cases? This modification is based on the papers of Hering (1950) and Jolivet (1954). The modified rules follow (see page 31).

Some examples of true phytophagous beetles are illustrated in plate 1 (p. 21) and plate 2 (p. 23).

Trophic Categories

Euphagy

The larvae or adults of these insects feed on the "normal" host plant or those considered as such. Thus, one can divide phytophagous insects into groups that are monophagous (sometimes called stenophagous), and euryphagous. The last category can be further divided into oligophagous and polyphagous, according to whether the animal feeds on several or many species of plants. This distinction, rather fine, is explained later, with examples taken from the Coleoptera, Lepidoptera, and other orders.

Monophagy

First degree monophagy The larvae or adults of insects that feed only on one species of plant are called strictly or specifically monophagous. Sometimes this is simply because some plant genera have only one species (monotypic), but more often species will be confined to a single polytypic genus, or even more frequently to a single family. Examples of this are rather rare among the beetles where many families, such as Meloidae (blister beetles), are polyphagous, but the Chrysomelidae are among the most specific in their food requirements.

In addition to the above mentioned case of *Papilio ulysses* in New Guinea that feeds only on one species of Rutaceous plant, *Evodia accedens*, we can cite several species of Diptera (species of Agromyzidae), such as members of the genus *Phytomyza*, miners in leaves of species of Umbelliferae and Ranunculaceae, and species of the lepidopteran genus *Mompha*, each of which are adapted to a different species of *Epilobium*. Other examples could be listed among the Lepidoptera, such as *Thyria celerio*, and others

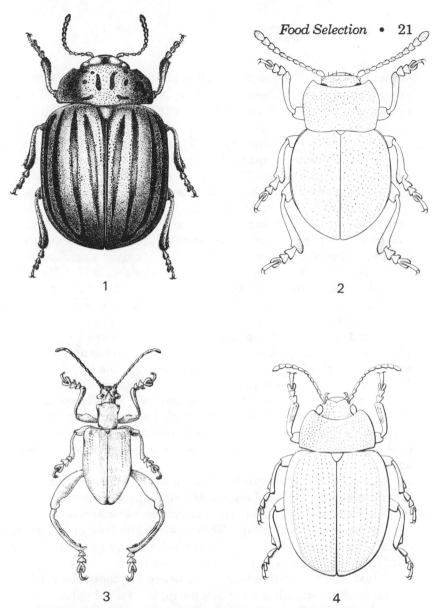

PLATE 1.True Phytophaga, some chrysomelid beetles. Fig. 1. The Colorado potato beetles, *Leptinotarsa decemlineata*, the most infamous among the Phytophagous beetles, which progressively, from a wild American species of *Solanum*, has invaded the North American and European cultivated potato patches. This is an example of restricted oligophagy. Fig. 2. *Timarcha metallica*, an European species linked trophically to species of Rubiaceae. It is a "bloody nose" beetle of the forest and mountains. Fig. 3. A species of Indo-Malaysian *Sagra*, probably *S. femorata*, gallicolous in the larval stage inside stems of leguminose plants. The whole subfamily Sagrinae has a Gondwanian distribution. Fig. 4. *Phaedonia inclusa*, a chrysomelid beetles which has become a pest of soy beans in Indonesia.

among the beetles (Chrysomelidae, Alticinae), that are well known and often are so specific that they are called biological races or even subspecies.

Second degree monophagy- This is where the larvae or the adults feed on several plants of a similar section of a plant genus. For instance, the miner fly, *Phytomyza abdominalis*, and the hymenopteran, *Pseudodineura mentiens*, live exclusively on species of *Anemones* (wind flowers) of the botanical section (or group) *Hepatica*. Sections among plant genera are natural divisions within the genus somewhat similar to the term subgenus as used in entomology. Other species of *Phytomyza* and *Pseudodineura* live exclusively on *Anemones*, section *Pulsatilla*; also several *Uredinea* (parasitic fungi, the rusts) of the genus *Puccinea* show the same selectivity on these plants. Similar examples are easily found among phytophagous Coleoptera, miners or not.

Third degree monophagy (or generic monophagy)- This is where the larvae or adults of insects feed on many or all the species of a given plant genus. For instance, the fly, *Agromyza nigrescens*, and the wasp, *Fenella voigti*, live on all the known species of *Geranium*. Numerous Lepidoptera, the Chrysomelidae (Coleoptera), which are the most selective among the phytophagous beetles, and others, may live on all the species of a given botanical genus, as for instance, the species of such genera as *Euphorbia, Rubus, Galium,* and *Solanum*.

Euryphagy or Oligophagy-Oliphagous insects (sometimes termed pleophagous) feed on plants belonging to several different genera. Here also, we can distinguish different categories.

Systematic oligophagy-This is when the food plant preferences of insects are confined to botanically related plants. We can easily apply Maulik's laws here (discussed later).

First degree oligophagy- These are phytophagous insects which feed on related plants belonging to several genera of the same family. For instance, the fly *Dizygomyza lamii* and *D. labiatarum* live on almost all the genera of Labiatae and *Philophylla heraclei* on almost all the Umbelliferae. That sort of oligophagy, common among Coleoptera, Lepidoptera, and phytophagous Hymenoptera, is not, however, to the same extent in all cases. For example, the European species of *Timarcha*, the black bloodynose beetles, in the European temperate climate, select most of the species of herbaceous Rubiaceae, and reject arborescent and exotic Rubiaceae. On the west coast of United States and in the

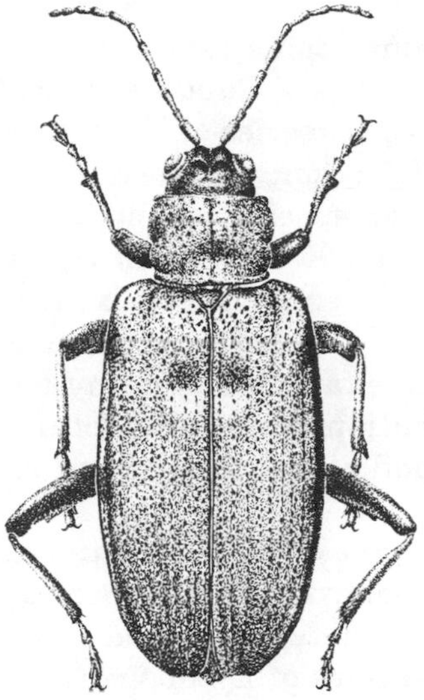

PLATE 2. True Phytophaga, additional species of chrysomelid beetles. Fig. 5. *Donaciasta garambana*, a semiaquatic and relatively polyphagous species. Fig. 6. *Platycornus peregrinus*, a beetle strictly associated (as are other species of the genus) with species of *Calotropis* (Asclepiadaceae). Many species of related genera feed on Asclepiadaceae and Apocynaceae. Fig. 7. *Chrysolina staphylea*, a species distributed in the cold areas of the Holarctic Region. Specimens are found in the subalpine area in South Korea where Labiatae and Plantaginaceae are their food choice. Fig. 8. *Nisotra gemella*, a flea beetle (Alticinae) from Thailand, feed exclusively on several malvaceous plants.

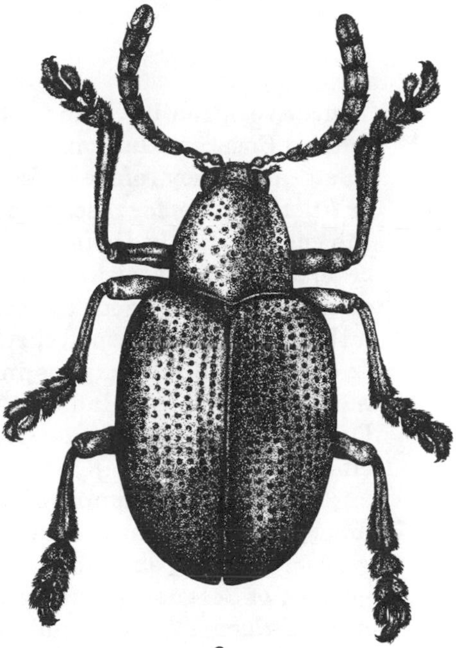

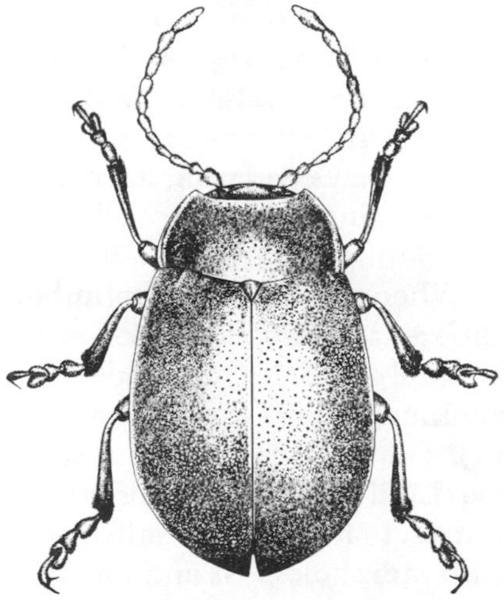

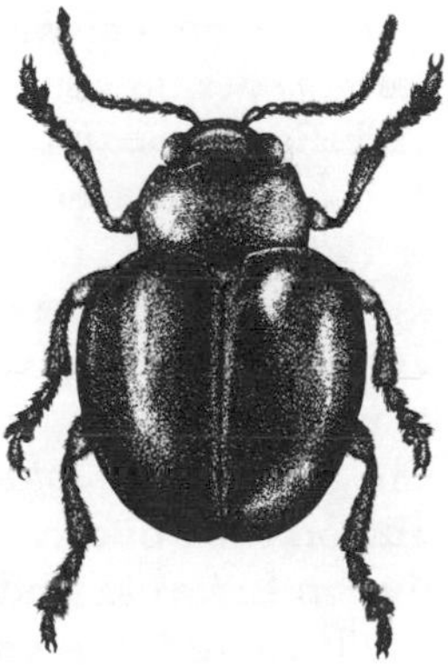

Vancouver area in Canada, the American species of *Timarcha* feed on Rosaceae, but only on the species of *Rubus* and *Fragaria*. In all cases, they refuse to feed on some species of *Galium, Rubia,* or *Rubus,* either for mechanical or for other, unknown reasons. In the case of the fly *Dizigomyza morio,* species feeding on Rubiaceae, as well as species of the European *Timarcha,* the rejection of some exotic Rubiaceous species may be explained by "xenophoby," a theory to be developed further on following pages. Among the first degree oligophagous insects, examples of many beetles, mostly Chrysomelidae and Curculionidae, Lepidoptera, and Diptera on Polygonaceae, Chenopodiaceae, and others, may be cited.

Still another example of first degree oligophagy is to be found among the phytophagous Acari (Eriophyidae). The species of the genera Vasates and Aceria are completely adapted to various species of Solanaceae, including species of *Lycopersicum, Solanum, Datura, Physalis,* and *Petunia.* Also parasitic fungi, such as *Peronospora parasitica* show this kind of oligophagy on Cruciferae.

Second degree oligophagy- Here the larvae and/or adults feed on plants of various genera belonging to related plants of the same order. For instance, the beetle, *Ceutorrhynchus contractus,* and the fly, *Scaptomyzella flava,* live on Cruciferae, Resedaceae, and Capparidaceae, all families belonging to the class Rhoedales or Papavelales.

Third degree oligophagy- When insects live on plants of a variety of genera, belonging to different orders, but which are still of related groups, this term may apply. Species of Lepidoptera of the genus *Elachista* live on Graminae, Cyperaceae (Glumiflorae) and on the Juncaceae (Liliiflorae). The fly, *Chrylizosoma vittatum* live on Liliaceae and Orchidaceae (Liliiflorae and Microspermae).

It is well known that some insects have a "botanical sense" which permits them to detect both morphological and physiological relationships that are totally unsuspected between plant families. Here also can be mentioned the theory of "bridge species" which are found both in the phytophagous insects and the parasitic fungi. According to this theory, two species which live on two different family groups of plants, not closely related, are able to transfer from one to another group through other plant species, the bridge species. These bridge species do not need to be related to either one of the families or group of families involved. There exists, however, the need of some common chemical substance, sometimes known to be essential for the two animal (or fungi) species. This is the "common denominator" proposed by Hering (1949).

From that common ground, each animal species or plant parasite, can migrate beyond its normal family group toward another species group. Among the bridge species, the most common is the garden nasturtium, *Tropaeolum majus* (Tropaeolaceae), a native of Peru, but widely cultivated in all parts of the world. The bridge effect works well for both the insect species and the parasitic fungi.

The rust, *Puccinia subnitens* (Uredinales), is found on one side of the leaves of Chenopodiaceae, Polygonaceae, and Amarantaceae, and on the other side on the leaves of the Cruciferae. *Tropaeolum majus* serves as a bridge species between these two groups of plants.

Many other examples, more or less complex, are known among the parasitic fungi and among the Coleoptera, Lepidoptera, and Diptera (see Jolivet, 1954).

Thus, according to the bridge species theory, if a species migrates to a species which provides common and agreeable chemical substances, it can then accept, as host plant, the other species by gradually becoming accustomed to the chemicals on the bridge species. However, such colonization is the fruit of slow evolution, often unfinished, in many insect families. The bridging substance contained, in the case of *Tropaeolum majus*, is myrosine, a diastasis which is also present among the Cruciferae. Mustard oil is found in many families of plants, among them, Resedaceae, Cruciferae, Tropaeolaceae, Capparidaceae, and some others, but it is totally absent from the Polygonaceae and Centrospermae. Therefore, it is hard to explain why certain Centrospermae feeding insects apparently utilize the bridge species *Tropaeolum majus*.

Combined oligophagy- Here belong phytophagous insects which may live on various genera of a single plant family, plus one species of plant of another family unrelated to the others. For example, the flea beetle, *Altica oleracea*, lives on Onagrariae and Lythrariae, related families, but also, this species is found on *Polygonum aviculare* (Polygonaceae), an unrelated plant. This may, however, be an ecological association because all of these are semiaquatic plants. Combined oligophagy, a term created by the Viennese entomologist, Heikertinger, a specialist on flea beetles, is also well known to mycologists. In these cases there cannot actually be any other satisfactory explanation. Here a subsidiary host is always an unique species, in contrast to the main hosts which are the species of an entire plant family. We have made similar observations among species of Altica in the tropics of Africa and Asia.

One might also cite the case of the Mediterranean species of *Timarcha* (Coleoptera) that have a rather complex food selection process, confined to Rubiaceae and related families, but also species of *Plantago* which many botanists consider to be unrelated to these families.

Many transitions exist between combined oligophagy and the following feeding habits.

Disjunctive oligophagy- The insects that live only on a limited number of unrelated plants of different orders are termed disjunctive oligophagous. Numerous flea beetles (Chrysomelidae, Alticinae), some Diptera, and mining larvae of Lepidoptera provide examples. For instance, *Liriomrza eupatoriae* (Diptera) lives on species of *Galeopsis* (Labiatae) and *Eupatorium* (Compositae). Also, *Lyonetia ledi* (Lepidoptera) lives on species of *Ledum* (Ericaceae) and on *Myrica* (Myricaceae). The last example could perhaps be explained because the two plants belong to the same plant association, but they certainly are unrelated. Disjunctive oligophagy is very rare. It may change to combined oligophagy, as in the case of *Liriomyza eupatoriae* which is rarely found on a species of *Solidago* (Compositae).

The term ecological selection, *i.e.*, selection of plants belonging to the same plant association, was used by Jolivet and Petitpierre (1979). As an example, the Australo-Papuan leaf beetles of the genus *Paropsis* and related genera are typical. In Australia they feed on species of *Acacia* (Leguminosae) and of *Eucalyptus* (Myrtaceae), families not related but growing together. Other examples may be found in aquatic, desert, grassland, and forest associations.

Temporary oligophagy- Oligophagy is said to be temporary, if, in its different stages of development, an insect species feeds in succession constantly on different and unrelated plants. This is the case of the insects with life cycles such as aphids or the rust fungi (Uredinales), and others. Among the leaf miners, one can cite several species of Lepidoptera of the genus *Coleophora* (*Lixella* group). At the first stage of their larvae these insects live exclusively inside the calyx or in the seeds of Labiatae, but late in the autumn or at the beginning of the following spring, they move to grasses of various species where they make large mines. It is difficult to explain the transfer from one family to another so different in their relationships (dicotyledons during their early stages and then on monocotyledons), a transfer probably made necessary by insufficient food during the long period of development.

Examples of temporary oligophagy among the Coleoptera are difficult to find. These exist, however, and when better known, the biology of floricolous and pollen eating beetles, such as species of *Orsodacne* and *Megascelis*, species of Oedemeridae, and in general, miners and gallicolous species, will probably offer more examples. Papilionidae (Lepidoptera) in the Middle East and elsewhere which feed successively on Umbelliferae and Rutaceae may be explained by chemical, but not botanical, relationships.

Many rusts produce two different types of spores during the duration of their cycle, each on different, unrelated plant families. These examples are so well known that it is hardly necessary to mention it here. There are similarities between the elaborate cycle of aphids and the rusts.

To explain this strange behavior, two hypotheses are equally acceptable: 1) original polyphagy with later specialized monophagy; 2) original monophagy which evolved later on to polyphagy on very different food plants.

Polyphagy

Polyphagous insects feed on a great number of plants belonging to unrelated genera in different orders.

First degree polyphagy- These insects feed almost at random on a wide variety of orders, but of the same class. Species of Lepidoptera of the genus *Cnephasia* live on various dicotyledons and seem to be attracted by herbaceous plants. Species of the genus *Phytomyza* (*P. atricornia*) and the genus *Liriomyza* (*L. strigta*), do the same thing. Among the beetles, Meloidae, and certain leaf beetles (Chrysomelidae), for example, *Galeruca tanaceti*, are equally polyphagous.

Second degree polyphagy- This is the utilization by an insect of plants of several classes (both mono- and dicotyledons). The moth, *Cnephasiella incertana* (Lepidoptera) and the weevil, *Orthochaetes insignis* (Coleoptera: Curculionidae), show this type of food selection, *Donacia* spp. among the leaf beetles.

An extended case of polyphagy is shown by the caterpillar

Amsacta meloneyi (Arctiidae), which appear in enormous numbers at the beginning of September in certain parts of the pre-Sahelian zone in Western Africa. However, it is not complete polyphagy. This hyperactive caterpillar covers hundreds of meters each day, devouring everything in its way, cultivated and wild plants alike, monocots, and even certain cryptogams. However, it rejects spicy plants such as red peppers and almost all the latex (milky sap) plants, Euphorbiaceae (*Euphorbia* spp.) and Asclepidaceae (*Calotropis* spp.). It is not surprising that it feeds on the species of the above families lacking bitter latex, but the vine, *Leptadenia hastata* (Asclepiadaceae) with translucent sap, also is eaten voraciously. This is interesting because species of *Calotropis* and *Leptadenia* (Asclepiadaceae), have similar attractants and are selected by Eumolpine beetles. Food selection in polyphagous insects, such as species of *Amsacta*, is based only on the lack of repellent chemicals. If *Amsacta* caterpillars eat *Catharanthus roseus*, the Madagascaran periwinkle (Apocynaceae, close to Asclepiadaceae), they reject absolutely the species with toxic latex such as those of *Adenium*. However, one cricket, *Zonocerus* sp., and others, eat and digest perfectly latex plants.

The weevils, among phytophagous Coleoptera, are very often polyphagous in the tropics. A good example in New Guinea are the species complexes of *Oribius*, and in the Mascarene Islands, the variable species of the genus *Cratopus*. Among the polyphagous animals that must be mentioned are Diplopoda (Myriapoda) which are linked mostly to hypogaeic habitats such as forest litter, under logs, under bark, or stones. Actually, species of diplopods have their eyes reduced or absent, more rigid exoskeletons, few spines, and other characteristics, while their carboniferous ancestors had larger eyes, bifurcated spines, and soft, flexible exoskeletons. There is much evidence that the fossil forms were external feeders, living openly on leaves in the forests.

Pantophagy

Pantophagous insects feed on almost all green plants and even on decaying vegetable matter. They accept as food, coniferous plants, angiosperms, ferns, mosses, and lichens. Fly species of the genus *Sciara* feed both on phanerogams (Compositae) and cryptogam species of *Marchantia*. This kind of selection also may be called third degree polyphagy. True panthophagy exists when insects accept dead and living vegetable matter equally. This is the case of clytrine larvae (Chrysomelidae) living deep inside ant hills where they feed on various plant detritus. They also feed on ant eggs and ant excreta. The adult on the other hand is phytophagous, feeding only on green higher plants. A true pantophagous insect is, for instance, the larvae of the Dipterous tendipedid miners, and Chironomidae, both of which feed on living plant material, animal and vegetable plankton, and even on decomposing detritus.

Xenophagy

Xenophagy (=allotrophy) pertains to a drastic change of the food habits of an animal brought about by extreme habitat stresses and is, as yet, relatively unknown. This should not be confused with temporary oligophagy, such as that found in certain hydrophilid water beetles, carnivorous in early larval stages, but later becoming herbivorous. Xenophagy is a complete change in diet, the herbivorous diet becoming carnivorous or vice versa, predators becoming necrophagous, coprophages becoming necrophagous or carnivorous, and so on. However, we deal here only with the change in the host plants among the phytophagous insects.

Xenophagy has its parallel among parasitic fungi (xenoparasitism). We have previously divided, rather arbitrarily (Jolivet, 1980), the known cases into experimental, forced, spontaneous, and fixed xenophagy. The last case is found among biological races not discernible externally, such as many flea beetles. Cases of forced allotrophy, *i.e.*, a drastic diet change in order to replace a lack of normal host plants, are very common. For example, *Papilio machaon* will eat Rutaceous plants, but dies on Umbelliferous plants in the Middle East, or *Vanessa io*, the peacock butterfly, when there is a lack of nettles (*Urtica* spp.), feeds on blueberries (*Vaccinium*). Many other examples of allotrophy are known. Certain cases are explained by chemical relationships, others remain unsolved (see Jolivet, 1954, and several other papers).

The butterflies of the genus *Papilio* and roughly speaking, all Papilionidae, may be divided into several groups according to their food choices. However, three groups are found rather characteristically on species of *Aristolochia*, on Rutaceae, and on Umbelliferae. The transition from Rutaceae to Umbelliferae, and *vice versa*, which rarely takes place, can be explained by chemical analogy (there are no taxonomic relationships), and bridge species containing anethol, such as the Rutaceous plant species of *Dictamnus* or *Pelea*. Although in countries such as Israel, the transfer is direct without bridge species. *Dictamnus albus* is often called the candle plant because of its strong, volatile, and inflammable ethereal oil which is secreted. It is this strong scent which attracts butterflies. This may have resulted in the formation of biological races. These races seem to be the result of a generally rare event, the choice of an "abnormal" host plant probably due to a sudden genetic change. Xenophagy in parasitic fungi occurs when the "abnormal" host plant is damaged or weakened, which in turn, seems to help the fungus colonize a new host.

Numerous leaf miners, such as those of Diptera and Lepidoptera, exhibit xenophagy. This phenomenon is not at all in contradiction to the "botanical sense" of the insects. Many "abnormal" mines come from the urgent need of the female to oviposit, and in the absence of the right chemical from the normal host plant, it lays its eggs on a new host plant. Sometimes the larvae are able to develop in that new situation. Caterpillars of the genus *Coleophora*, namely *C. fuscedinella*, have been observed mining flowers of species of *Caltha* (Ranunculaceae). The larvae had evidently fallen down from an overhanging tree which was their ordinary host. The same species was once found mining a species of *Lysimachia* (Primulaceae) below a birch tree (*Betula* sp.). Similarly, *Coleophora flavipennella* has been found mining *Helianthemum* sp. (Cistineae) below an oak, its normal host plant.

Among the reasons for such mistakes, smell must be very important, the "abnormal" host plant odor often being masked to the egg laying female by the smell of the normal food plant (a Labiate, for instance). That evidently is the case when two plants, the usual and the abnormal hosts, coexist in the same habitat.

In order for a new host plant to be acceptable, the insect must find among the attractive substances (olfactive and gustative), at least one that is not also repulsive, and it must not be exposed to any toxic compound within the plant. All of these conditions are rarely met.

Many other factors seem responsible for the xenophagy of the miners. The greater succulence or juiciness of the leaves in cultivated species, the higher temperature of greenhouses, and even walls exposed to the sun, all bring about biological disturbances which disrupt the normal biological equilibrium. Also, it must be stressed that, besides the real cases of xenotrophy, some may be cases of oligophagy, which may be distinguished only after a deeper knowledge of the natural relationships of the botanical families (for example, Amentaceae and Rosaceae).

"A vegetable pill,
Were I to swallow this," he said,
"I should be very ill."
-*Lewis Carroll: Sylvia and Bruno.*

Chapter 3

Problems of Food Selection

If we go deeper into the study of food selection of phytophagous insects, we find that other problems exist which have not yet been approached and that this selection is not as simple as one might believe. Various ideas introduced by several authors (for example, Hering [1950] and Maulik, [1947]) referred to previously, need to be discussed in more detail.

Xenophoby

Concepts- Xenophoby is the total rejection by many insects, even those showing a third degree monophagy or an oligophagy, of any foreign plant, even those very closely related to its usual host. But there are degrees of rejection. In certain cases, an insect accepts the foreign plant, but is unable to complete its life cycle if it feeds on it. Some insects accept imported plants close to their normal host plant (Xenophily) and other systematically refuse them. Xenophoby in phytophagous insects is more common than is generally believed. The problem of food choice seems to be related to the geographical origin of the plants. These stages and examples of each follow.

Examples - Many rusts (Uredinales) show, beyond any doubt, xenophoby. Among insects, a good example is the Colorado potato beetle, *Leptinotarsa decemlineata*. It accepts species of American Solanum, but in Europe, it eventually, but not easily accepts *S. dulcamara*, and totally rejects *S. nigrum* and most other Solanaceous genera. When the plant is not totally rejected, the whole cycle of the insect is disturbed, but the exact reason for this has not been determined.

Most of the European and American Coleoptera, Lepidoptera, and Diptera specialized for feeding on species of the Polygonaceae genera (*Rumex, Polygonum,* and *Oxyria*) die on the leaves of rhubarb (*Rheum Rhabarbarum*) which is of Manchurian origin. On the contrary, *Gastrophysa atrocyanea,* a small blue beetle living in Japan, Korea, Eastern China, and North Vietnam, accepts the Asiatic rhubarbs, but evidently prefers species of *Rumex.* The oligophagous fly, *Pegomyia hyoscyami,* attacks only Central European Solanaceae and rejects the potatoes of American origin. The caterpillars of the moth *Lyonetia clerkella* which live on Central European species of *Betula, Castanea, Salix, Pyrus, Malus,* and *Prunus* (Amentiferae and Rosaceae), which is a rather large trophic spectrum, rejects, on the other hand, the American *Prunus,* or if it is accepted, the foreign plant kills the insect. A similar case of lethal foreign oligophagy is known among the moths, *Gracilaria syringella,* on species of *Chionanthus* (Oleaceae), although it normally accepts a great variety of local Oleaceae as well as some Caprifoliaceae. [Many species of Chrysomelid beetles are strictly phytophagous. Several of these species are illustrated in plates 1 and 2, and are referred to in various parts of the text.]

Examples of Xenophily‒ Conversely, xenophily is the acceptance by an insect of an introduced foreign plant closely related to the normal host. Xenophily, very different from xenophagy (allotrophy), is less common than xenophoby. Xenophagy may be total, or partial in the Colorado potato beetle, which accepts in Europe, as mentioned previously, *Solanum dulcamara,* but rejects *Solanum nigrum.* The most striking cases are those insects associated with species of *Lonicera* (Caprifoliaceae), the honeysuckle, which also accept, in general, species of *Symphoricarpus,* (also Caprifoliaceae). This shrub, which is becoming established in Europe, originated in Canada and was introduced around the year 1817 in various parts of the world. Species of moths of the genera *Limenitis, Diselachista,* and *Lithocollis,* and species of flies of the genera *Phytomyza* and *Phytagromyza* accept this introduced plant without any difficulty.

Cause‒ (of Xenophoby) The reason for the development of lethal monophagy and oligophagy are rather difficult to understand. In order for an insect to successfully feed on a given plant, there are several essential conditions to fulfill: 1) a rather large ecological preadaptation, 2) an attractant substance present, 3) agreeable taste, and 4) absence of any repulsive or toxic substance which can inhibit the development of the insect. It seems that the geographic isolation of plant species between the Nearctic and the Palearctic regions since the Eocene has brought with it a divergence of proteins and even some heterosides. However, the problem is not clear. It seems difficult to imagine that, in the previously cited case of *Lyonetia* sp., the differences between the European and American *Prunus* spp. are greater than those between species of *Prunus* and species of *Betula*. Hering (1949) tried experimentally to prove, through tests in the Botanical Garden in Berlin, his theory, which states that the number of plants attacked by leaf mining insects decreases proportionally to the distance from the original locus of the plants. In other words, plants from distant localities have much less chance of being eaten than plants from close by localities. If confirmed, the principle could help to determine the autochthonous origin, or not, of a given plant species. Probably related to this phenomenon, equally inexplicable but not unique, is the case of the pink hawk moth, *Hyloicus ligustri*. This moth lives, as larvae, indiscriminately on lilac (*Syringa vulgaris*), privet (*Ligustrum vulgare*), and ash (*Fraxinus excelsior*), all plants of the Oleaceae (olive family). Indeed, the moth shows rather pronounced oligophagy. If, however, we raise caterpillars from eggs on lilac or privet, it is then impossible to change the food plant from one to the other during their development. The caterpillars will start to eat, but soon die. Some kind of adaptation to a toxic chemical must take place only in the early stages of the larvae, and because of differences between the toxic substances in the two plants, the larvae do not survive the change.

Maulik's Law

Maulik, the late chrysomelid specialist, in 1947 formulated a general law which fits phytophagous insects as well as parasitic fungi, provided they show monophagy, or at least systematic oligophagy. Maulik's law, as it is now called, is only a simple deduction and an expression of a well known fact:

a) If a phytophagous insect accepts, as food, a whole group of plants belonging to one or more genera, another insect that accepts one of those plants will also accept all the plants of the group.

This law has very few exceptions. The main one is xenophagy (allotrophy). Therefore, its corollary is equally true:

b) When a species of a group of insects has been found feeding on a plant, one can expect that other insects of the group will also accept the plant.

The second rule is not universal, but it is a good basis for prediction. For instance, when a new species of *Lema* is described in the Old World, one can expect, without too much chance of error, that its host plants will be species of Commelinaceae, Orchidaceae, Dioscoraceae, and related plants, or for species of *Crioceris*, Liliaceae seem the best host plant to be expected, and for other species of *Lema*, Solanaceae or Gramineae. It is evident that the first law is supported by numerous examples chosen among Coleoptera, Diptera, Tenthredinidae (Hymenoptera), and Lepidoptera. For example, the relationships between the plant groups *Rumex-Polygonum*; *Lonicera-Symphoricarpus*; Cruciferae-Resedaceae-Tropaeolaceae-Capparidaceae; Vitidaceae-Onagrariae-Lythariae; *Sarothamnus-Cytisus*, and others as cohosts, have been described.

Although all insects that live on Amentiferae also accept Rosaceae, which seems to be a good indicator of the probable relationship between the two plant groups, this relationship is not yet admitted by the botanists. Another explanation could be ecological. The "ecological selection" mentioned previously dealt with the host plants *Acacia* spp. (Leguminosae) and *Eucalyptus* spp. (Myrtaceae) in Australia. Also in temperate climates, the forest is mainly composed of Amentiferae, and the scrub areas of Rosaceae. Ecological selection would be an easy way for insects to solve their food plant selection problem, especially those insects that are relatively polyphagous. The herbivorous layer, mainly composed of Gramineae and Rosaceae, is rarely reached by these insects.

Maulik's law evidently has a biochemical basis and the attractive principles of the chosen foodplants are supposed to be found in each of the potentially acceptable alternate hosts.

Botanical Sense (of insects)

The expression, botanical sense of insects, originated with the nineteenth century French entomologist, Jean Henri Fabre. He applied the term to indicate that sort of almost infallible instinct with which insects are able to recognize plants taxonomically related, but often morphologically different. That observation is good only for second and third degree monophagy and oligophagy. It is evident that the chemical affinities detected by insects are not always visible externally and botanists should sometimes take into account the choice of phytophagous arthropods, mostly in cases of little known exotic species of plants.

The value of the botanical sense must not be exaggerated; however, it is often useful. While the presence of oligophagous species on two given plants may be interpreted as indicating a relationship between plants, their absence cannot mean a lack of relationship. The same rules may be applied to host choice of parasitic fungi on plants. Also, it is evident that this peculiarity can help in identifying leaf mining species forming new mines according to the kind of plant attacked. Also it should be noted that even if the relationships between several plants is clearly suspected by entomologists, there is no absolute proof of that relationship. The final classification is that of the botanist. The host plant selection of species of *Papilio* that make ecological selections between Umbelliferae and Rutaceae do not follow the rule of "botanical sense" at all. The best and most typical example of botanical sense is to be seen in the moth genus *Thyridia* (Lepidoptera, Neotropidae) of the Neotropical region. One species of caterpillar of that genus feeds on species of the genus *Brunfelsia* which formerly was included by botanists in the Scrophulariaceae. However, all the other species of the genus *Thyridia* feed on Solanaceae. Then it was discovered that the caterpillars were consistent in their food choice and that the genus *Brunfelsia* really belongs in the Solanaceae.

The moth, *Euspilapteryx phasianipennella* lives on Chenopodium, *Atriplex* (Chenopodiaceae) and *Polygonum* (Polygonaceae). Both plant families are closely related. The moth also lives on species of *Lysimachia* (Primulaceae), which, according to the old classification, is only distantly related to these two plant families. It is now admitted, however, that there exists a close relationship between Primulaceae, Polygonaceae, and Chenopodiaceae. On the contrary, the old floras showed a close relationship between *Plantago*, and *Galium*, but new classifications widely separate these two genera. We know that *Timarca* spp., the black bloodynose beetle in Europe and North Africa, often feed on species of both genera of these plants. Hence, perhaps the old botanists were right and both families may have to be carefully reviewed. Leaf mining species have shown a close relationship between the genera *Angelica* and *Arcnangelica*, so close, in fact, that the two genera are now merged.

On the contrary, various species of leaf miners and also the parasitic fungus, *Gymnosporangium sabinae*, prove that the separation between *Pyrus* and *Malus,* often criticized, is perfectly valid.

We may conclude, therefore, that the selection of food by phytophagous insects and fungi can sometimes be a valuable help to the botanist. Again, in the final analysis, the taxonomists must decide.

Hybridization and Chimaeras

We have now reached the problem of the colonization by an insect of plants similar to (xenophily) or different from (xenophagy) those normal for the species. The conditions governing this colonization are dealt with in the following chapter on the choice mechanisms. Here is mentioned only the effect of hybridization on monophagy or oligophagy.

Plant hybrids or Chimaeras- Plant hybrids, even chimaeras, formed from a kind of mosaic of tissues from different species in one plant, produced artificially by grafting, can transform a first or second degree monophagy into third degree monophagy or into systematic oligophagy. It seems, for instance, that *Nepticula nanivora* (Lepidoptera) which is restricted to *Betula nana,* the dwarf birch, evolved from *Nepticula betulicola,* a species mining *Betula pendula* and *B. pubescens.* It is supposed that *B. nana* has been colonized by the hybrid of *Betula pubescens* and *B. nana* which is very common in the same area occupied by *N. nanivora.* This same hypothesis seems to be a correct explanation of several allotrophy cases found in species of *Lithocolletis* (Lepidoptera). New species may well be produced this way.

It seems true that monophagy also may be transformed into systematic oligophagy thanks to chimaeras or graft hybrids. It is well known that those composite plants which can be reproduced only by cutting are made of two kinds of tissues of two different genotypes, generally a wrapping tissue of one species and a wrapped tissue of another species. Species of *Crataegomespilus,* for instance, are made of an external tissue of *Mespilus* sp. containing *Cretaegus* sp. tissues or *vice versa.* In Berlin, according to Hering, only leaf miners of species of *Crataegus* were found on the chimaera (species of *Nepticula, Parornix, etc.*). No doubt those chimaeras, which are much more widely distributed than is ordinarily believed, contribute to the extension of monophagous species' diets.

Animal hybrids- Hydridization between two animal phytophagous species or subspecies (sometimes called races) can, for some unknown reason, so deeply modify the hybrid metabolism that the offspring can accept a host plant rejected by the parents. The hawk moth, *Celerio euphorbiae* (Lepidoptera), is represented by two races, *C. e. euphorbiae* and *C. e. mauritanica*, both of which feed only on species of euphorbias and not on species of willows (*Salix*). The hybrid, *C. mauritanica x euphorbia* (=*wagneri*), and the reverse, can be fed entirely and without any problem on species of willow, plants not related to the Euphorbiaceae. The same result is obtained with the hybrid *C. euphorbiae x galii* (=*kinderwateri*). Its caterpillars willingly eat species of *Salix*, plants rejected by both parents. As yet, these facts have not been explained.

Numerous examples of behavior disturbance among hybrids could be cited to support these previous examples. For instance, the male of the hawk moth, *Pergesa elpenor*, that feeds on species of *Epilobium*, has been crossed with the female of *Celerio hippophaes* that eats only *Hippophae rhamnoides*. The females normally lay eggs on *H. rhamnoides*, but the larvae resulting from this cross accept only *Epilobium* spp.. Several crossings between males of the geometrid moths that live on various deciduous tree leaves and females that feed on *Achillea millefolium*, an herbaceous plant from the under story layer, give hybrid larvae that accept only the food plant of the male. Many examples are known among the Lepidoptera and there must be more among other phytophagous insects.

Butterfly memory

Whether or not butterflies have a memory is a much debated question, mostly in recent seminars on food plant selection. We shall say only a few things on this subject, which otherwise would necessitate a chapter of its own. This deals with the return of butterflies and moths, when egg laying, to the previous food plant which fed them when they were caterpillars. It is, if fully confirmed, a very remarkable adaption. The adult butterfly, when egg laying, becomes sensitized to the color green. Previously, when collecting nectar or pollen, only bright colors like red, orange, and blue interested it. However, mostly the olfactory sense functions as the main selective factor for the mature female.

The classical example of butterfly or moth "memory" is shown by *Hyponomeuta padellus* which consists of two biological races, one on apple trees (species of *Malus*) and another on hawthorn (species of *Cretaegus*), two closely related plants. The race living on hawthorn, even if fed experimentally on apple trees, produce female moths that prefer to lay eggs on hawthorn and *vice versa*. The food change has not at all modified the ancestral food habits of the moth.

There can exist among oligophagous and even polyphagous species, some rather well marked food preferences. This is a good example of the famous Hopkin's Principle, often discussed and frequently questioned. According to this principle, the adult insect of oligophagous and polyphagous species has a tendency to lay eggs on the host plant species which was its food during the larval stage. This is not in contradiction with the previous case of *Hyponomeuta podellus*. The former deals with biological races; here we deal only with a large spectrum of differences within the same species. Alternative host plants chosen by an insect generally contain the same attractive substances. However, a subtle secondary substance, mustard oil, attracts *Pieris brassicae* and *P. rapae*, cabbage butterflies, to species of Cruciferae, Resedaceae, Tropaeolaceae, and Capparidacea. Under the name mustard oil are grouped various sulfuric esters (alkyl isothiocyanates), slightly different from one another, found in species of *Brassica, Reseda*, and *Cochlearia*. These slight differences can result in a rather strict choice of host plant by oligophagous species. Also, certainly many insects, chiefly species of Lepidoptera, have slight chemical differences which deserve to be genetically screened for possible sibling species.

It is evident that the psychological phenomenon involved (memory) is a bit anthropomorphic. A butterfly cannot remember the host plant which fed it when in the caterpillar stage. Probably sometimes, and for certain species, a genetic selection of ultra selective lines during breeding takes place and that selection is made at the cost of different tastes. However, the olfactory reactions of species of *Drosophila* suggest associative learning (Atkins, 1980). For the adult, peppermint oil is normally a repellent, but if the larvae are reared in the presence of the chemical, the resulting adults are attracted to it. That can be a key to Hopkin's principle's applicability to leaf eating species where the secondary substance differs from plant to plant.

Dethier and Yost (1979) have clearly shown, in experimental tests, that the caterpillars of the tobacco moth, *Manduca sexta*, do not learn to reject a plant which previously made them sick. They have been fed on *Atropa belladona* or insecticide treated *Nerium oleander*, or on a species of petunia. The larvae, after recuperation, accept without hesitation the previous plants and do not learn from the experiment. This seems to be totally different in the case of vertebrates (which demonstrate bait rejection, or the Garcia effect) and at least some invertebrates such as slugs. Food aversion learning is well known among rodents. We admit that rats have a better memory than insects, but slugs with a better memory seem strange. This would seem to refute the hypothesis of the induction of food habits among the invertebrates, at least among the oligophagous species. Polyphagous invertebrates differ in this behavior, at least among insects and mollusks.

Resistant plants

When agriculture developed in the fertile Eastern crescent along the Mediterranean Sea, in Peru, in Mexico, in Thailand, and elsewhere in the world during the post glacial period, humans began to slowly improve the cereals, consciously or not, by selecting the biggest seeds (corn, wheat, barley, rice, sorghum, oats, rye), and the ones resistant to diseases and pests! This selection at the dawn of agriculture was made almost blindly and only with certain species, but it resulted finally in the composition of our present day cereals. Continued selection of commercially produced seeds has further improved resistance. Research to produce insect and disease resistant plants has long been a main objective of agronomists and geneticists.

In an effort to combat the diseases and insect pests of food crops, researchers have applied these methods in the development of biological control programs. Such programs use natural predators, parasites, and parasitoids of insect pests and plants resistant to disease, insects, and mites, in conjunction with cultural techniques, genetic control, and biotechnical methods (employing hormones, pheromones, repellents, attractants, *etc.*). Finally, through integrated control methods (proper doses of insecticides combined with biological control techniques), an attempt has been made to devise a balanced system with the goal of improving the agroecosystem. The use of plants resistant to insects and disease is a method which by itself has often given excellent results. Chrysomelid beetle preferences for poplar tree (*Populus* spp.) clones have been studied. Mangoes resistant to scale insects and fungi have been developed and propagated. Many other examples might be cited. Unfortunately, the best varieties are not always the resistant ones and sometimes the wild species are more tolerant of pests and disease than the cultivated varieties. Certain plants, for example the gymnosperm, *Ginkgo biloba*, the Maidenhair Tree, still growing wild in the Chekiang Mountains in China, have no, or very few, pests outside its country of origin. The same thing is true for the species of eucalyptus which, due to the repellent effect of their essential oils, do not incur any leaf miner attacks outside the Australo-Papuan area, and species of *Hevea* (rubber trees) are practically uninjured by pests in Africa. In the latter case, the plants seem resistant not only because of insect "xenophobia," but also because they are toxic and repulsive to attackers. Of course, Asclepiadaceae and Euphorbiaceae, the latex plants, have many parasites and pests perfectly adapted to their high toxicity, and resistant varieties of cultivated plants are neither more nor less toxic than the normal varieties. They show only slight differences in chemistry or morphology but this still makes them rather safe from predator and parasite attacks.

One can also believe that certain plants can repel insect attacks, not only with toxins, but also with hormone mimics like the growth inhibitors found in coniferous trees and ferns. As early as 1951, Painter published a book on resistant plants. Since then many books and articles have been published on the subject. It is impossible to list them all here. We can cite such works are Beck (1965), Van Emden (1966-1973), de Wilde, *et al.* (1969), and the recent review on the subject in Metcalf, *et al.* (1975). Research continues on an international scale.

Hybridization, which can extend, as we have seen before, monophagy and oligophagy, can also produce the converse, *i.e.*, plant hybrids resistant to insect and parasite attack. Research also has been oriented towards clonal varieties by systematically testing resistance and selection of spontaneous and induced mutation.

One well known example of resistance to insects is the case of certain species of *Solanum* and the Colorado potato beetle, *Leptinotarsa decemlineata*. This problem has been the object of research for more than 80 years and is still without any clear cut results. The species of *Solanum* have been classified into four sections according to their susceptibility to the Colorado potato beetle: 1) highly susceptible species, 2) moderately susceptible species, 3) moderately resistant species, and 4) resistant species. The cultivated potato belongs to category 2. It is not the most favored plant by the beetle. It was not the original host plant since the edible potato originated in the mountainous areas of South America. It is also evident that mechanical obstacles (hairs in this case) are often as repulsive as chemical obstacles. Also, there is no relationship between the species attractiveness for egg laying and for larval feeding. Several species of *Solanum* neglected by the adult Colorado potato beetle are sometimes favorable for larval development and vice versa. This is, to some extent, due to the morphological adaptation of the larval legs to the host plant (Jacques and Arnett, *in litt*).

The aim in the struggle against the Colorado potato beetle is to obtain, by a clever crossing of a species of *Solanum*, a hybrid which retains the food qualities of the potato tuber and possesses the noxous properties of other species such as *S. demissum*. It is impossible to maintain a permanent colony of the Colorado potato beetle on *Solanum demissum* or one of its hybrids with the tuber potato, *S. tuberosum*. The reason seems complex. *S. demissum* does not allow the larvae to feed properly and also it produces a repellent alkaloid, demissine. Most of the hybrids obtained between the potato and *S. demissum* and other American species show variable resistance to the Colorado potato beetle because the hybrid sometimes is recessive. On the other hand, and this is the problem, all of those hybrids do not produce tubers similar to the ones produced by the cultivated potato, and this is only part of the story. Forty years after the beginning of research on this subject, the problem is still far from being solved.

Research on resistant plants is approached from two different directions: accidental discovery of a population of resistant varieties, or systematic research among the available varieties. A wild species of *Solanum* has been crossed with a potato utilizing colchicin and compatible chromosomal formulas, thereby, resulting in a hybrid species. By this method it has been possible to produce aphid resistant potatoes.

The definition given by Kogan (1975) of plant resistance to insects can be stated as follows: it is that property which permits a plant to ward off, tolerate, or recuperate from the attacks of insect populations which would cause much more damage to other plants of the same species, under similar environmental conditions. This property is essentially biochemical or morphological as we have seen above, and is also a latent preadaptation within the species genes, as it is in insecticide resistance among insects. A classical example is the African cassava, where the bitter varieties with high starch content are protected from predators by the presence of cyanogenetic glycosides.

Among insects, in addition to the classical resistance against species of *Phylloxera*, pests of grapes, excellent results have been obtained by the discovery of resistant alfalfa, barley, beans, maize, sorghum, rice, sugar cane, and corn. Through research we are now quickly developing resistant cotton, pumpkins, onions, peanuts, sweet peppers, potatoes, soybeans, tobacco, several grains (*e.g.*, oats), and other species.

Several timber trees are naturally resistant to termites. Since Kogan's study, although rather recent, even more studies have appeared. Kogan distinguished in plants an ecological and genetic resistance, essentially either chemical or morphological. The influence of environmental factors on the expression of that resistance should not be neglected.

"Of the botanists, the most dangerous species,
The one who introduces...."

"A Naturalist."

Chapter 4

Biological Weed Control

The entomological literature of the world is full of descriptions of host plant specificity of phytophagous insects (and sometimes also fungi) used in the fight against imported weeds which have become pests in their new country. We refer to the classical entomological handbooks for more details about these examples. Actually, there exist American entomological teams based in Rome, Italy, and Australian ones (CSIRO) in Montpellier, France, for the purpose of conducting research directed toward the biological control of weed species of the genera *Carduus* and *Chondrilla* (Compositae), imported into the United States of America and Australia respectively, and many other plants such as species of *Heliotropium*. The work of the Australians consists of identifying, in the Mediterranean region, the plant and animal parasites of the dandelion, *Chondrilla jacea*. Then, these are tested in the laboratory on various ornamental and useful plants cultivated in Australia, and even on wild species such as eucalyptus trees that belong to an unrelated family (Myrtaceae). When the tests are positive, *i.e.*, if one is certain the pathogens and parasites are selective, then only with the greatest precautions, are these species cultivated. After a reasonable quarantine, they are released in their adoptive country. Sometimes, in places such as Australia where the quarantine laws are very strict, some insects (*e.g.*, coprophagous beetles) are imported only in the egg stage which have been externally sterilized.

The Australian region has proved an ideal experimental area for this kind of research and numerous other countries have been encouraged to begin research, and consequently to introduce auxiliary phytophagous animals for plant control. We can state, for example, that the Hawaiian Islands, India, Mauritius, Madagascar, South Africa, New Zealand, continental United States of America, and Canada have specialized teams, most often composed of an entomologist and a phytopathologist, for these studies. The phytopathologist is in charge of studying the viral, mycoplasmal, bacterial, and fungal diseases of the plants to be controlled.

Even if the fight against some invading plants has been a success, there is still a problem with certain others. Some of those for which control is being studied are: *Convolvulus* spp., *Senecio jacobae, Rubus* spp., *Cyperus* spp., *Xanthium* spp., *Kentrophyllum lanatum, Centaurea calcitropa, Echium vulgare, Matricaria discoidea, Salidago* spp., *Galinsoga* spp., and particularly the famous water hyacinth, *Eichornia crassipes.* The latter is now being controlled by the weevil, *Neochetina eichorniae* in southern Florida.

It is much easier to fight against invading plants than against insect pests, provided one takes into account a series of essential rules. The most important rule is to insure the nonpathogenicity of the introduced species against useful indigenous species.

One of the best successes realized so far, is the control of the gorse, or whin, *Ulex europaeus,* by *Apion ulicis* in New Zealand, Southern Australia, and Tasmania. (It seems odd that this biological control measures have never been attempted in the "Hauts" of the island of La Reunion, near Madagascar. This plant, introduced by a nostalgic Briton two centuries ago, is becoming a pest at the expense of grazing land on that island.) The partial destruction of lantana bushes (in Hawaii and Fiji) with a moth (Tortricidae), a fly (Agromyzidae), another fly (Trypetidae) and a bug (Tingidae) is another example of, at least, limited success in the use of insects as control agents. Rounding out the list of successes are: the destruction of the St. John's wort, *Hypericum perforatum*, mostly thanks to several species of *Chrysolina* (Chrysomelidae), in Australia and California; the control of *Clidenia hirta* with the thysanopteran, *Liothrips wrichi*, in Fiji, and the control of the prickly pear cactus (*Opuntia* spp.) with the help of a moth, *Cactoblastis cactorum* (Crambriidae) and several species of cochineal scale insects in Australia, India and South Africa.

Among the successful application of biological control is the control in Hawaii of the weeds *Eupatorium adenophorum, Tribulus* spp., *Emex australis, Hypericum perforatum,* and in Russia, orobanche, a parasitic plant on sunflower roots, controlled by the

fly, *Phytomya orobanchiae*. Orobanche does not have any green
tissue and, like mistletoe, is hemiparasitic. Some select specific
hosts such as tomatoes in the Sudan, while others infect many
species. It is impossible to control these pests with chemicals
alone. Similarly, *Cuscuta* spp., the dodders, cause damage to
trees and irrigated crops in the semidesert steppes of Afghanistan. So far, in those areas, no biological control has been tried.
Strangely enough, the dodderlike *Cassytha filiformis* (Lauraceae),
of the Old World tropics, does not seem to parasitize cultivated
plants, or if it does, it has never been reported.

The problem sometimes is more complicated. The control of
the prickly pear cactus *(Opuntia* spp.) is aimed at the destruction
of the species with spines and not the spineless ones which can be
used to feed cattle (although in Mexico the spines of all species
are burned off making it possible for them to be used as cattle
feed). In certain areas, namely in Madagascar, there has been
some success in eradicating the prickly species without affecting
the spineless ones, but specificity of an insect to one of such
similar species is extremely rare. In South Africa both kinds have
been destroyed altogether.

It should be noted that invading plants in the tropics are
often ubiquitous, having been carried by humans wherever they
have gone. Many weeds originated in tropical America or Southern Europe. Personally, I have found, in such far away places as
Sahel (an area in west Africa at the southern limit of the Sahara),
New Guinea, La Reunion, and Thailand, the weeds *Euphorbia
hirta, Sida rhombifolia, Tridax procumbens,* and many others. In
the highlands of New Guinea I have been able to follow the
progression of the European *Plantago major* along the roads and
in the villages. This large plantain was introduced, via Australia,
and surely is filling an empty ecological niche. It replaces at a
middle elevation, the endemic woolly plantains of higher altitudes. Some introduced plants are relatively easy to fight using
food selection specificity and introducing parasites and pathogens from the native habitat of the weeds.

Control experiments began in 1950, in Mauritius, against the
Central American *Cordia macrostachia* (Bignoniaceae), a small,
rather common, arborescent shrub. Two chrysomelid beetles were
introduced, one, the cassidine *Physonota alutacea,* and the other,
a galerucid, *Schematiza cordiae.* Both are active tree defoliaters
in Trinidad. Twenty five years later, while visiting in Mauritius, I
observed that the weed had persisted but was integrated into the

local flora and was practically under control. However, only the galerucid beetle had survived, the cassidine did not become established.

During the preliminary studies in preparation for the introduction of a phytophagous parasite, as we have seen previously, the species must be imported free of all its enemies. Its selectivity has to be carefully studied to determine that it is harmless to cultivated plants, and it must be determined that it can adapt to the new habitat. The fecundity of the species also must be determined. One must be very careful about possible xenotrophic (allotrophic) changes and mutations which can change its behavior and sometimes produce polyspecific and oligophagous varieties.

When it was decided to destroy the blackberry, *Rubus fructicosus*, in New Zealand, someone suggested the use of the buprestid beetle, *Coraebus rubi*, native to southern France. The idea was abandoned because this oligophagous insect is perfectly able to attack roses and other cultivated Rosacease. The cure would have been worse than the ill. The case of the "vigne marrone" *(Rubus moluccanus)* ,on the island of La Reunion, is interesting. Introduced 200 years ago by a missionary from Indochina, the plant rapidly became a pest, particularly in the mountains where its edible berry is disseminated by birds. In this case, biological control theoretically appears simple. One, after following strict screening procedures, could use the native phytophagous enemies from Vietnam or Thailand, such as the chrysomelid beetles, *Phaedon fulvescens, Chlamisus latiusculus,* and other species of *Chlamisus,* several eumolpines such as species of *Basilepta,* a certain number of moth caterpillars and a fungus (rust). They all seem to be host specific and are not pests of roses. Additionally, there is a virus which probably is transmitted by aphids. More studies are needed in the native land of the plant and its country of adoption before releasing this virus.

Unfortunately, insect introductions are numerous and although some of them, such as species of *Phylloxera,* and the Colorado potato beetle, were accidental, many others were deliberate. The introduction of *Papilio demodocus* on citrus in La Reunion and Mauritius is an example, even when endemic species of *Papilio* existed in the mountains of both islands.

The island of La Reunion, where deliberately introduced weeds such as *Agave, Rubus moluccanus, Eugenia jambos, Lantana, Ulex, Solanum auriculatum, Erigeron* spp., *Fuschia* spp., and a great many others are numerous, now has hectares of grassland and arable land covered by these weeds. Except in Mauritius, almost nothing has been tried to control the weeds, even though something could be done to regain this valuable land.

The method used for selecting beneficial insects consists of selecting monophagous species and rejecting oligophagous and polyphagous species. On the other hand, it is rare that only one phytophagous species is sufficient to destroy an invading weed. Some cases exist, however, as for example, the control of *Opuntia* spp., and *Hypericum perforatum*, but even with these two, their insect enemy has been helped to some extent, by some other species. Generally one selects a whole series of pathogens and phytophagous arthropods which will attack the leaves, flowers, seeds, stems, roots, and fruit. Their combined effects may succeed in controlling the weed.

The rare failures are due to the presence of either parasites, or predators, on the phytophagous control agents. The fact is well known that, in Australia, a parasitoid of *Paropsis* sp. attacked, after a relatively long period, *Chrysolina* sp. which were imported to fight *Hypericum perforatum* (St. John's wort). Most often nonselective predators are present which rapidly adapt themselves to a new prey. Another reasons reason for failure is the change of climate due to the seasonal inversion between the two hemispheres. Generally, however, that obstacle is rapidly corrected by both plants and animals. Seasonal differences sometimes increase or diminish the number of generations of the introduced insect. *Chrysolina varians*, for instance, has doubled the number of its generations per year since its introduction to Australia.

In conclusion, let us emphasize the importance of having exact knowledge about the life cycle of phytophagous insects. No detail should be considered unimportant.

Elton (1958) lists three stages at which plant and insect invaders may be repelled. First, we can prevent them at the gate by quarantine. Second, we can destroy the first bridgehead by eradication. This is a very rare phenomenon. It was a success in Brazil for certain mosquitoes and in Bougainville, in 1969, for the Giant African Snail, *Achatina fulica*. Finally, and this is the most reasonable method, we can keep the invader population below the dangerous threshold via biological controls. Fighting against noxious weeds using phytophagous control agents is one means of control which renders an introduced plant rare and harmless.

Provided that rigorous screening procedures are followed, success is much easier to achieve when controlling weeds than fighting insect pests.

"De gustibus et coloribus
non est disputandum"

-The Scholastics.

Chapter 5

How Food Selection Works

A great deal of research during the past seventy years has concentrated on the physiology and chemistry of food selection. As is quoted above, "everyone to his tastes and colors." Findings have been published in many journals and have been discussed in many symposia. Insects in these experiments belong to various orders, but primarily beetles and moths are used. Leaf mining flies do not adapt well to such research and their manipulation is difficult. Only in one rare case has a leaf miner been utilized, that is the leaf mining beetle larvae, *Coelaenomenodera* sp., cultivated in the laboratory only for biological control experiments. The Colorado potato beetle, *Leptinotarsa decemlineata,* and various other leaf beetles with short life cycles, many generations per year, and a facultative diapause, such as species of *Gastrophysa,* have provided first rate research material. The production of synthetic food media for phytophagous insects has been an aid in rearing them and hence, in the understanding of the mechanisms involved in plant selection.

We should not forget that although insects have the senses of sight, touch, taste, hearing, and smell, they are different from ourselves. These senses may be acute, sometimes even more than ours, but their sense organs are different. The receptors for taste, for instance, are located on their palpi, and often, especially among flies, beetles, and butterflies, taste is also perceived via ventral pads on the tarsi. In Lepidoptera, the tarsal receptors inform the insect about the exact place of food on the substrate after smell has brought them from a distance. We refer the reader to the data on the physiological chemistry of insects for a detailed study of these senses. The spatial location of the receptor organs does not change the nature of food choice.

Food selection depends on sight, smell, and taste. The sense of touch sometimes plays an equal role with the other senses. For

example, repulsive properties of prickly plants with hairs and
thorns are determined by touch in certain kinds of phytophagous
insects. Secondary substances, either toxic or maybe only variations within a given plant species, can be a reason for selection
(or repulsion) in cases of "xenophobia," at least according to
Hering (1951). Mechanical selection through the sense of touch
certainly plays a role in the local selection of certain parts of the
host plant. When an insect accepts a new plant it must be preadapted to it. Hence we are going to distinguish the following types
of trophic selection:

Mechanical or physical selection

A monophagous insect can refuse a plant with the same
chemical compounds, or the same attractive substances as its
normal host plant only because the toughness of the leaves, the
presence of hairs, or a cuticle renders it difficult to bite. Insects
that eat species of *Galium*, such as species of *Timarcha* in continental Europe, and the Galerucine, *Sermylassa halensis*, reject
Galium aparine and *Rubia peregrina*, as too hard and too hairy,
at least among the old plants. It is a fact, often observed in
Europe, Asia, and the U.S.A., that species of *Chrysomela* are
found only on the young growth or shoots of willow and poplar
trees. Very often, we have observed in the Austral-Papuan region,
species of *Paropsis*, big goldenred chrysomelids, on young shoots
of eucalyptus, but never on old leaves.

The little cabbage diamond moth, *Plutella xylostella*, lays eggs
more easily on a rough, unequal surface than on a smooth one.
Mustard oil is one of the selective factors for moths feeding on
cruciferous plants. The caterpillar of *Lasiocampa quercus* can
feed on the hard surface of the leaves of the holly tree, *Ilex* spp.,
only if the dentate edges are cut off, since it always eats the leaf
edges. The bean jassid, *Empoasca fabae*, is disturbed by the
density of hooked hairs (trichomes) of certain varieties of beans.
An aphid pest, *Aphis craccivora*, reacts in exactly the same
manner. Resins (on coniferous trees), siliciums, or oxalates of
other plants, provide certain plants with resistance to pests.

Species of moths of the genus *Lithocolletis* are host specific
on young oak shoots, on young birch, or *Scabiosa* sp. (Dipsaca-
ceae) depending on the species. It is also evident that the chemical composition of a leaf varies with the age of the plant, shady or
sunny conditions, time of day, and so on. Two insects that feed on
the same plant, one during the day, the other during the night,

may be considered to have, in a sense, two ecologically different diets.

On the contrary, the caterpillars of *Acronycta abscondita* prefer stunted individual plants of *Rumac acetosella* and *Euphorbia cyparissias*. Such cases are not at all isolated. Species of caterpillars of the genus *Acidalia* eat only wilted leaves and some *Pterophora* spp. larvae cut certain veins causing the leaf, that they are going to eat later, to wilt. It must be noted, large polyphagous larvae, like the caterpillar of the African *Amsacta meloneyi* (Arctiidae) that eat practically everything, prefer fresh leaves and do not attack wilted ones. However, this selection is not purely mechanical. Chemical factors in the composition of the leaf are also factors in food selection. Other examples will be discussed further on.

Selection of feeding sites

Selection of a feeding site limited to a specific part of a plant is made especially among leaf miners, but aphids also display area selection when sucking plant juices. Leaf mining larvae choose, depending on the species, the cells of one of the two epidermal layers of a leaf, either the palisade tissue, or the palisade and parenchyma tissues. The first selection is often due to the difference between the chemical composition (protids) of the different layers. If the egg is laid on the wrong side of the leaf, the larva often refuses to feed and perishes. Attacks by leaf miners can also modify the metabolism of the leaf and, consequently, the nature of the products ingested. The most striking example is that caused by the larvae of the moth, *Nepticula basalella*. The leaves mined by these caterpillars remain green in the feeding area a long time after the departure of the pest and even after the leaves fall to the ground.

Among the sap sucking insects (leafhoppers, aphids, reduvids, and others) and root eaters (eumolpines, elaterids, and so on), the choice of plant tissue is very limited. That choice seems to be based on the sense of taste, or perhaps more precisely, on the ability to distinguish various pH gradients. Very little is known about the way insects with ovipositors select their host tissue. In gallicolous species, for instance, this selection is very specific. It is possible that chemoreceptors provide the insect with all required information.

Selection is less restrictive for caterpillars, but it does exist. Large caterpillars eat all of the lateral part of the leaf to the

stalk, but the small sized caterpillars often start at the apex of
the leaf where the veins are less developed. Other caterpillars
skeletonize the leaf, leaving the veins, while others eat only the
upper or lower surface.

The number of examples of localized food selection could be
multiplied. The method of selection is eminently variable in the
same insect and on different plants. Some local selection is more
mechanical, some more chemical, or a combination of both of
these.

When insects are adapted to a part of the plant (root, stem,
flowers, fruits, seeds, or leaves), knowing that their selection site
can be mechanical, chemical or tropistic, we can experimentally
alter the selection of the part it will use, but the insect generally
makes this change very reluctantly. The Colorado potato beetle
will feed on potato tubers and even on other parts of the plants in
addition to the leaves which comprise the normal diet, but only if
forced. Some insects, for example, the moth, *Heliothis* sp., that
easily accept a purely synthetic food, do so only when forced,
because the normal attractants, such as the green color, the
smell, and the tissue texture, are absent. The caterpillar of
Phalonia ambiguella which lives during the spring on grape
flowers and other plants, and during its second generation on the
fruit of the same plants, shows some adaptation to the host's
development (which is an example of coevolution). *Larentia incul-
traria* does the same with leaves and capsules of *Primula* spp..

One could cite certain gallicolous cynipids also, such as
Biorrhiza pallida, that show an alternation of generations on an
oak *(Quercus* sp.), one generation gallicolous on the aerial parts of
the tree, the other gallicolous only on the roots.

As is mentioned by Dethier (1953), the chemical composition
of the green plant varies with the time of day, the season, the
stage of development of the plant, the tissues fed upon, the climate of the environment, the nature of the soil, and also the
insecticides and fertilizers used to cultivate the plant. The chemistry of the plant is not uniform and permanent for all individuals
of the species, or its varieties. Consequently, this explains, in part
at least, the local or seasonal behavior of insects and also the
ability of certain plants to resist certain insect species. Schoonhoven (1972) writes that many insects can evaluate the nutritive
composition of their food and choose accordingly. Those insects,
even if theoretically polyphagous, are in fact selective.

Visual selection

It is a well known fact, as previously mentioned, that butterflies, during the oviposition or egg laying process, cease to be interested in brightly colored flowers and become interested only in the green leaves of their larval host plant. During this period, they rest on green or bluegreen surfaces and actively move their anterior legs, probably to get olfactive and gustative perceptions of the substratum via the tarsal receptors. Some moths, such as *Manduca sexta* (Sphingidae), cannot find their host plant if the antennae are cut, and that action, therefore, totally prevents oviposition. This detection is first visual, and then olfactory. A third factor, taste, may intervene. The females of the cabbage butterfly, *Pieris brassicae,* and the diamond moth, *Plutella xylostella,* cruciferous feeders exclusively, will lay eggs on any surface provided the anterior legs stay in contact with a cabbage leaf, or even some mustard oil on filter paper. Some aphids are attracted by the yellow color of the aged leaves. Among the saturnids and several other nocturnal species of Lepidoptera, the perception of the host plant (oak, for example) seems to be chiefly olfactive and localized on the basiconic sensillae of the female antenna.

It is wrong, however, to say that all caterpillars make no choice for themselves and that the adult chooses the plant for them. Certain butterflies and moths deposit their eggs at random when flying near the host plant without alighting on it. In that case, the caterpillars must make a food choice. This will be dealt with later on. Many caterpillars, such as those of *Heliothis armigera,* are evidently attracted by green, and an observer crossing a tomato or cotton field can find them on their clothing if the cloth is green.

Visual selection, therefore, even if not always preponderant, plays a certain role in food choice. Many chrysomelid beetles, often brightly colored themselves, distinguish color very well. Species of *Chrysolina* can separate violet from green, blue from yellow, and orange from violet. Certain chrysomelids are even more selective. They are capable of distinguishing the various tones of pale green, but not dark green. These insects probably use this ability in their host plant choice. Recent experiments with the watercress leaf beetle, *Phaedon cochleariae,* have shown that at least one phytophagous species has a specific reaction to yellow and green wave lengths, and that this is clearly associated with host plant selection. However, color itself is not the only criterion of selection and the role of tint and light intensity must not be overlooked.

The conclusions drawn from the research on the watercress leaf beetle have shown clearly that this insect is attracted to its cruciferous host plant by the yellow and green colors, and also by the odor of mustard oil. Mustard oil odor provokes feeding and this act continues, induced by plant glucosides. Leaf texture also affects its feeding, but experiments show that sight and smell are the chief elements of food selection at least at the beginning.

It is known that some caterpillars distinguish three main colors, blue, green, and white. Color also plays significant role in the choice of flowers by their adults, especially for honey gathering activities, as well as during the choice of the host plant where it will deposit its eggs. For instance, species of *M. acroglossum,* day flying hawk moths (Sphingidae), mainly visit blue flowers when young and later, when nearly ready to oviposit, they prefer yellow-green plants and flowers. Yellow-green is the color of *Galium* spp. (Rubiaceae), which feeds the hawk moth caterpillars.

These reactions are not isolated facts. Even night flying moths are sensitive to color, only in a different way. We have seen that the hawk moths in New Guinea visit only white, smelly rhododendrons, while their odorless, red flowers were only visited by birds. In that case, brightly colored flowers are invisible to these nocturnal species, which however, could see the pale colors. Bees are also able to detect various colors, but in a different part of the color spectrum. This fact is too well known to describe in detail here.

Aphids lay eggs mostly on yellow-green or orange surfaces, provided the colors are not saturated, but the colors chosen are often in relation to the specific host characteristics. Yellow pan traps are classical means for attracting and capturing these insects.

Phytophagous beetles, such as Chrysomelidae and some Coccinellidae, as well as carnivorous ladybird beetles, will follow a dark band painted on white paper and turn when the band changes its direction. This tendency seems to be linked with the habit of climbing up stems and branches to reach leaves. Caterpillars show a crude perception of shapes. This perception, even though imperfect, helps them to find the host plant when they have dropped to the ground, but it is also associated with chemical senses which separate the shady forms they see. From past research, it appears that the nature of the stem is not enough to provoke the caterpillar into climbing. It is also necessary that at the top there be a screen which casts shade on the ground. Of course, the caterpillar cannot appreciate, with its small ocelli, a primitive sight organ, the shape of the leaves.

Wandering, solitary caterpillars such as those of *Amsacta meloneyi*, which in Africa devour everything in their way, certainly must have a more acute sight mechanism than that of sedentary caterpillars. The larvae of certain beetles also are attracted by vertical forms.

A good example of the influence of the shape of the host is cited by Kogan (1976) for the American moth, *Autographa precationis* (Noctuidae), which generally does not lay its eggs on the dandelion, *Taraxacum officinale*, the favorite food of its larva. It seems that this is due to the low habitus of the plant. The adult prefers the soybean's tall and vertical habitus, which means that the larvae must migrate to the proper host.

Certain Neotropical butterflies of the genus *Heliconius* are very interesting because their caterpillars feed on leaves of the passion flower, *Passiflora* spp. (Passifloraceae). These butterflies have excellent vision and do associate shape of leaf with food, even though these leaves vary enormously among the 350 species of the genus *Passiflora*. Strangely enough, the caterpillar is associated with species of passion flowers, but the adults visit flowers of species of *Anguria and Gurania* (Cucurbitaceae). It is very rare to find similar host plant specificity among adult butterflies (Gilbert, 1975). Another peculiarity of the selection of passion flower vines by the females of species of *Heliconius* has been recently discovered by Gilbert (1982). It seems that in defense the vines have evolved fake eggs that make it look to the butterflies as if eggs have already been laid on them. This acts as a repellent to the ovipositing female.

With the visual senses are associated various tactile receptions: photo-, geo-, anemo- and hydrotropisms. These factors, together with chemical factors, play an important role in the choice of food and oviposition. However, the main determinants in food selection are the chemical factors discussed and which are further described in the following paragraphs.

Chemical selection

Chemical selection of the host plant by insects operates through both the sense of smell and taste. The presence or absence of poisonous chemical substances in plant tissues plays an important role. For example, the extremity of the rostrum of many Hemiptera, such as those of pyrrhocorid bugs, is provided with basiconic sensillae, chemoreceptors responsible for food choice. Simply stated, in addition to the above mentioned factors, for a plant to be acceptable, it is necessary that: a) the plant produces an attractive smell, b) it does not produce a repulsive smell, c) its taste is agreeable, and d) it does not contain any product which is poisonous, sterilizing, or inhibiting to larval or pupal development. Of course, the last condition becomes a selection factor only after the fact. Normally an insect will accept those plants which later induce sterilization, or inhibit growth, provided the attractants, either smell or taste, are present. This is, however, rather rare in nature because most adaptations have been fixed for a long time. It usually occurs only in the laboratory or in imported plants. Chemical selection is less evident among aquatic insects where taste and smell are combined since all those stimuli must be mixed together in a solution.

Thorsteinson (1960) used a formula to summarize food selection. This can be stated as follows:

$$F=-I-D+Esn \ [Ep]$$

An optimal feeding response (F) exists when the substratum is devoid of inhibitors (I) and of repellents (D) and contains the necessary chemotactic stimuli (E) in order to encourage feeding. Stimuli, E, are sapid products (Esn) and sometimes special kinds of "pungent" stimuli (Ep), which are facultative. The secondary substances belong more to Ep, however. This formula requires taste, but does not mention the sense of smell which is often equally important for attraction and the continuation of feeding.

Locust food selection has been studied in the Cape Verde Islands. What appears to be a mistake of instinct is cited for the locust (grasshopper), *Pyrgomorpha cognata* and the sacred tree, *Datura innoxia*. In these islands, food is scarce, and choice is restricted further at the end of the rainy season. *D. innoxia* is a plant known for its high level of poisonous alkaloids within its leaves. Sometimes in Cape Verde, female grasshoppers have been observed eating the leaves of *D. innoxia*. Dead or dying locusts and abnormal egg laying, have been seen on the leaves as well, which for this species, is completely aberrant. This plant, if eaten, seems to be extremely toxic for the insects and inhibits reproduction (Duranton, *et al.*, 1982). Similarly, the widely polyphagous caterpillar of *Heliothis armigera*, which normally feeds on tomatoes, cotton, beans, maize, *etc.*, accepts, in Cape Verde, cabbage leaves, nontoxic, but unusual, perhaps because of the scarcity of available food.

Generally butterfly adults discriminate their host plants better than their caterpillars do. However, the silk moth, *Bombyx mori*, which has lost not only its flight abilities, but also all selective sense, lays eggs anywhere and its caterpillars accept many unrelated plants.

Various species within a natural family of plants generally show a remarkable similarity in chemical composition (*e.g.*, Umbelliferae with essential oils; Solonaceae with alkaloids; Cruciferae with myrosin or mustard oil; Polygonaceae with oxalates; Rubiaceae with glucosides; broom with spartein, and St. John's wort with hypericin; Cucurbitaceae with cucurbitacins; beans and other legumes with cyanogenetic glucosids; Rutacease with various essential oils, and many more). These substances are mostly selected by taste, but some of them also have a fragrant odor (*e.g.*, Rutaceae). Volatile substances such as terpenes found in trees also attract scolytids and other ambrosia beetles.

Cruciferous feeding insects, Lepidoptera, Coleoptera, and aphids, for example, are attracted by mustard oils and their glucosids. Mustard oil also attracts small braconid wasps (Hymenoptera) that are parasitic on many phytophagous insects. These substances also stimulate spore germination of the parasitic fungi, *Plasmodiophora* spp., on cruciferous plants. It must be noted, however, that the oviposition preferences of adult insects seem to be narrower than the choice capacities of the larvae. Since 1910, researchers have tried many experiments to demonstrate this difference. The earliest attempts were made by Verchaffelt (1910) using the white cabbage butterflies, *Pieris brassicae* and *P. rapae*, which feed exclusively on Cruciferae, Capparidaceae, Tropaeolaceae, Salvadoraceae, and Resedaceae. All of these plants have special glucosides (isothiocyanates) in common. Verchaffelt succeeded in getting the caterpillars to feed on plant leaves of different families, such as Leguminosae, by impregnating the leaves with diluted mustard oil. It must be noted also that even if the pierids feed mostly on Cruciferae and related families, in certain areas of the world species of the genus *Cassia* (Leguminosae) had been chosen by several species of this family, namely species of *Catopsilla* and *Callidryas*. This is an excellent example of combined oligophagy, perhaps due to mutation.

Similar experiments have been tried with the host specific leaf beetle, *Gastrophysa viridula,* and with other phytophagous insects with the same results.

The role of both a;la;pods amd chlorophyll in food selection cannot be neglected. Experiments using synthetic media made from these materials have shown that even if a polyphagous species, a *Heliothis,* for example, willingly accepts a relatively simple medium; oligophagous species are much more difficult to feed and require a more carefully compounded synthetic medium.

The role of smell also seems predominant over taste for food selection, since in the dark caterpillars rarely make mistakes, even without previous gustative trials. The sense of taste is, nonetheless, important. For a long time, it has been thought that this importance was negligible since the caterpillar could not distinguish the four taste qualities (sweet, sour, bitter, and salty), but only whether the food was agreeable or unpleasant. Taste distinction is, on the contrary, very well developed in some adult butterflies which are not only nectariphagous insects, but sometimes accept pollen, putrid food, fruits, urine, or excreta as food. I remember when I was in Kivu, many years ago, a lepidopterist kept human excreta in his refrigerator and offered it every morning in his garden, placing it on a papaw leaf to attract the splendid species of *Charaxes* butterflies. Many species of Nymphalidae and Papilionidae, for example, *Papilio weskei,* are strongly attracted by fruit, rum, and even human urine.

Schoonhoven (1968), analyzed the reactions of chemoreceptor cells of oligophagous caterpillars to distinguish between the receptors sensitive to sugar (glucose, sucrose, or both), salt, alkaloids, glucosides, inositol, amino acids, and anthocyanins. His results indicated a greater complexity than previously thought.

Schoonhoven's studies of food selection among phytophagous insects shows three selection stages: 1) discovery of the plant, 2) gustative trail, and 3) consumption. These studies, made with more than 20 different caterpillars, have clearly shown that each species has a characteristic receptive system and no two are identical. Hence, this explains trophic preferences and specializations. A monophagous species would differ from a polyphagous species because it tolerates fewer secondary substances.

Chemosensitive phenomena certainly govern the host plant selection of St. John's wort by species of *Chrysolina,* or for the mint chrysomelid beetles (species of *Chrysolina, Cassida*), and *Solanum* by *Leptinotarsa* spp. Rees (1969) has identified chemoreceptors and mechanoreceptors (trichoid sensillia) on the fifth tarsomeres of the legs of these species. After their arrival on the host plant, apple tree aphids detect a flavonoid (phloricin) which is found on the leaves and the stems of the tree. This is probably the substance which attracts the insect. That locusts (grasshoppers) nibble a plant before laying their eggs is well known, and the process, in a relatively eclectic species, shows a certain discriminating choice.

Nutritive substances used by sucking insects seem to be similar to the ones eaten by other insects. Aphids can be bred easily on synthetic nutrient media. Synthetic media have shown clearly the effect of the quality of the food on the morphological expression of aphid polymorphism, itself also genetically controlled. The age of the plant also has a physiological effect on the insect as de Wilde (1969) has proven with the Colorado potato beetle.

Species of Orthoptera, which are almost all polyphagous, bite into plants without prior chemical stimuli and try all at random. Taste also prompts the choice of the breeding place, since the adult feeds on the same plant as the nymphs. Species of Tettigonidae chew the plant before laying eggs. Despite the polyphagous nature of the Orthoptera, certain plants with latex, such as species of *Calotropis* (Asclepiadaceae) used as food by certain species, are rejected by others. Aphids also try their food by short and repeated insertions of their proboscis into different places on the plant or on different plants. Outside the olfactory and gustatory senses, a rather vague chemical sense must also exist among insects. In fact, the location of these sense organs is variable and not well known.

According to certain authors, the proteins, amino acids, carbohydrates, fats, sterols, mineral salts, and the B group vitamins content of green plants are relatively constant for all plants, though toxic substances do vary. Plant fats are not necessary for insect nutrition.

Aphids and mites attack plants only during certain seasons, even in the tropics. It seems that acidity of sap varies which may provide temporary resistance for the plant. No formal proof shows that seasonal changes in plants are responsible for migrations to different host plants for alternation of generations in aphids. Specificity of host plants could be due to substances such as alkaloids, cyanogenetic glycosids, flavonoids, terpenoids, essential oils, saponins, tannins, and acids acting as chemical stimuli, either positive or negative. These are the secondary substances of various authors. However, Hering (1949), formerly believed that proteins were responsible for that selection, but this theory was discarded after the present biochemical findings.

The sense of smell, that is to say, the reactions of the olfactive apparatus to a specific stimulus, is generally more developed in the adult insects than in their larvae. Research on the Colorado potato beetle's smell perception has been conducted in various countries. It has been shown in the past that the distance from which the larvae of that beetle can see the host plant is about 2 mm. Thus it seems clear that the perception of the host by a larva, which is practically blind, must be primarily olfactive. The antennae also play a major role along with the maxillary and labial palpi. Ocellae are involved in phototropism which together constitute the perception of the light intensity. Water vapor tension, the nature (texture) of the substratum, and the physiological conditions of the larvae also take part in chemotropism.

The gustative sense seems to be localized for the Colorado potato beetle larvae mostly in the mouth cavity, which differs from that of flies or butterflies, for instance. On the internal surface of the labrum or the labium of the beetle larvae are found placoid and trichoid sensillae, the gustative receptors. There are tangoreceptors (touch) diffused more or less over the entire body. Finally, the antennae and the palpi contain basiconic and styloconic sensillae in definite numbers and positions. According to certain authors, these are olfactive receptors, but only because no liquid is involved.

Schoonhoven (1974) also mentions sensillae on the galea, maxillary, and labial palpi of the Colorado potato beetle larvae and those sensillae are certainly taste receptors and phagostimulants. The tarsal hairs of *Leptinotarsa descemlineata* contain receptors for sugar, salts, and alkaloids. One of the St. John's wort leaf beetles, *Chrysolina brunsvicensis,* has tarsal chemoreceptors which are sensitive only to hypericin. Other receptors are stimulated by salts, but not to sugars. Those leaf beetles which do not eat *Hypericum* spp. do not react to hypericin, which surely indicates that it is the selective secondary substance. Sensillae of the maxillae of species of *Pieris* caterpillars are sensitive to mustard oil, and to hypericin in the beetle *Chrysolina brunsvicensis.* This could also be one of the mechanisms for detecting specific attractants. Schoonhoven also shows that the majority of the chemoreceptive cells of the oligophagous Lepidoptera are localized among the styloconic sensillae of the maxillae.

The consumption of leaf matter by a phytophagous insect varies in quantity in proportion to its size. For example, studies of a small species of American leaf beetle, *Systena blanda*, show that one individual eats about 0.3 cm of beet leaves per day. Neither density nor temperature modify this process. It is evident that large species such as the Colorado potato beetle eat proportionally much more.

As far as we can judge, all leaves contain in quantity the nutritive substances needed for growth and development of insects. Therefore, the big question is, what makes the differences between plants that attract some insects more than others. These differences are subtle. Odor, taste, toxins, and nutritive substances are some of the factors responsible for their choice. Essential oils and alkaloids are responsible for the taste and smell of the plant.

When an insect attacks a plant, it is equally true that it is necessary for the growing period of the plant to coincide with the feeding stage of the insect, or *vice versa*. If those conditions are not fulfilled it is what Painter (1936) called "evasion or pseudoresistance." This peculiarity is utilized in devising cultural methods and a simple difference in time between date of sowing seed and hatching of insect eggs can provide crop protection.

Hsiao (1969), studying the Colorado potato beetle and the alfalfa weevil, found that chemical stimuli influencing host searching are the following: attractants, repellents, stimulating signals, feeding, and rejecting stimuli.

In summary, the conclusions reached by Painter, Dethier, and others, is that sight, phototropism, geotropism, anemotropism, and hydrotropism play a significant role in insect orientation for both oviposition and feeding, and that the elements which attract at short distances are essentially chemical. Insect behavioral activity, therefore, is as follows: a) orientation toward food; b) response to biting or sucking; and c) continued feeding.

It is not necessarily the same smell which stimulates orientation that stimulates feeding, as has been proved in tests using elaterids (click beetles). Kogan (1975), summarizing numerous experiments, describes the selection processes for phytophagous insects as being a chain of events between the plant stimuli and the responses given by the insect. According to him, five stages can be distinguished: a) finding the host plant in its habitat; b) discovery of the individual host plant; c) recognition of the host plant; d) acceptance of the plant; and e) chemical acceptability as food of the host.

At phase d) several chemical substances seem to control

various phases of the feeding process. For the silkworm, a series of compounds existing in the mulberry leaf are associated with the first bite, then with ingestion, and finally with increased feeding. These stimulants are chiefly secondary metabolic compounds exuded by the plant epidermis and provide the stimulation necessary to trigger all of these processes.

Recent research, namely electrophysiological, on food perception among insects has demonstrated that odor and taste seem intermixed and are decoded inside the central nervous system as "host" and "nonhost." Some recent classifications have correlated the stimuli produced by the plants and the insect responses during the selective process. Those secondary chemical factors of the plants have been termed "allelochemicals" by Whittaker and Feeny (1971). They define these as the "chemical products by which the organisms of one species affect the growth, health, behavior, or biology of an organism of another species, food excluded." Such a relationship can exist between plants and insects. Pheromones (ectohormones) are essentially chemical messages between the individuals of a single species and are very different from the "allelochemicals" which affect individuals or populations of a species different from the one producing the chemical. Also involved are allomones, chemical agents of adaptive value for the producing organisms, and kairomones which have an adaptive value for the receptive organisms. Table I is slightly modified from Kogan (1976), Dethier (1960), Beck (1965), and Whittaker (1971), and shows the distinctions between the main groups of chemical factors and their effects.

Secondary substances produced by the plant, in principle responsible for the trophic selection by insects, theoretically are toxic toward other animals or plants or at least repulsive to them. This includes biochemical protection (allelopathy) of the plant against pathogens, herbivores, or competition by other plants. They have been named "allelopathic" substances. There are, however, cases, as mentioned by Whittaker and Feeny (1971), where the hormones, pheromones, or the nutritional compounds produced by a species also react as kairomones for another species.

Certain insects succeed in avoiding toxic or repulsive substances by various means, including the consumption of withered leaves devoid of these materials (certain aphids or Microlepidoptera) or the selection of nontoxic parts (sap sucking Hemiptera and Homoptera). Also it should be noted that aphids prefer young

or old leaves to the mature ones and on these they are more pro-
lific. Though often polyphagous and with alternate cycles, aphids
clearly show botanical preferences rather difficult to analyze. The
choice of young or old leaves is probably linked to their cyclic
development.

TABLE I

Main groups of chemical factors produced by plants (allelochemicals) and the corresponding behavior or physiological effect
on insects.

Allelochemicals	Behavioral or physiological effects
A) Allomones (repellents)	Gives an adaptative advantage to the producing organism (plant)
Repellents	Orient the insect outside the plant.
Locomotor stimuli	Start or accelerate movement.
Suppresents	Stop biting or peircing
Preventents	Prevents continuation of feeding and oviposition.
Antibiotics (growth inhibitors)	Prevent normal growth and larval development; reduces longevity and fecundity of the adult.
B) Kairomones (attractants)	Gives an adaptative advantage to the receiving organism (insect)
Attractants	Orients insects toward their host plant.
Arrestants	Slows down and stops movement.
Excitants	Produces biting, piercing, or oviposition.
Nutritional stimuli	Provokes continuation of feeding.

An interesting note about *Leptinotarsa decemlineata*, the
Colorado potato beetle, is that without its normal host plant and

the potato (*Solanum tuberosum*) itself, we have seen it accept or reject many species of the genus *Solanum*. Normally the insect is attracted by species of Solanaceae, but according to certain authors it can also live experimentally (not in nature) on Compositae (for example, *Lacturasativa*) and Asclepiadaceae (for example, *Asclepias syriaca*). The first plant is a "neutral" one, but the second is highly toxic, as are all the plants of that family. Perhaps there is some kind of preadaptation in the genus *Leptinotarsa* similar to other genera, for example, *Labidomera* which develops normally on species of *Asclepias*. Arnett (*in litt.*), however, has shown that *Leptinotarsa lineolata* is very host specific on *Hymenoclea monogyra* in Arizona and cannot be forced to eat other plants even those mechanically acceptable to the larvae. In any case, that shows that the natural and normal preference for the Solanaceae is not due totally to nutritive substances, but to a specific attractant. The researches of Fraenkel show that strong trophic selection is directed by the presence or absence of secondary chemical substances and not by nutritious substances.

Hsiao (1978) has shown that the Colorado potato beetle is an indigenous oligophagous species in North America and may adapt to ten indigenous Solanaceae other than the natural host, *Solanum rostratum*. Races, varieties, or biotypes may actually develop and form those oligophagous species. The geographical isolation of the insect and the host plant is surely a very important element. Hsiao insists that this is different than intraspecific variation in the case of the polyphagous species where the formation of biotypes is adapted to the host and is produced sympatrically. According to Hsiao, beetle populations from North America show a great deal more interpopulation variations as compared to European beetle populations which were introduced accidentally in 1922. According to Mitchell (1982), host plant selection by Colorado potato beetles appears to be largely mediated by alkaloids found in Solanaceous plants, and feeding deterrents caused by steroidal alkaloids. Recent taxonomic researches on the genus *Leptinotarsa* by Jacques (1989) completely challenges the long criticized, classical research of Tower. During our research on chrysomelid host plants, in collaboration with E. Petitpierre, we have found a very clear evolution of certain oligophagous genera from a basic, primitive choice (for example, *Timarcha* spp. on *Galium* to species of *Plantago* and others; species of *Chrysolina* from Labiatae to a certain number of very different plants). A common denominator exists between species of *Timarcha* and

Chrysolina. It is *Plantago* spp., but never has a species of *Chryso-lina* been found feeding on species of *Galium* or a species of *Timarcha* on Labiatae. Experiments could be attempted, at least with a species such as *Chrysolina staphylea* which is potentially oligophagous. However, very often the common denominator is missing and probably extinct, having developed along several lines during the course of evolution. We have also found a certain relationship existing between the chromosome formula of species of *Chrysolina* and their trophic preferences, which can only confirm the taxonomical relationships between their several ecologically distinct groups. As for the evolution of all these related genera of chrysomelids, this can be explained by an ancestral polyphagy, and a subsequent specialization, or by a deviation from a basic monophagy on Rubiaceae, Labiatae, Salicinae, Myrtaceae, *etc.*, with botanical relationships easily explaining the insect's choice. Other preferences remain unexplained, as for instance, the fondness of willow and alder insects for Rosaceae (*e.g.*, species of *Agelastica,* and others).

It is also well known that toxic secondary substances, which protect the plant, later protect monophagous insects which feed on it. Numerous are the known cases, as for example, toxic Altici-nae from Kalahari desert which are used by the bushmen for making arrow poison; danaid butterflies and species of *Poekilo-cerus* locusts on species of *Calatropis,* the small red moth, *Euche-lia jacobaeae,* on *Senecio jacobaeae*; zygaenid moths and their host plants, and many others. In the case of the Kalahari flea beetles, the food chain is rather complex, because their mimetic carabid beetles, the Lebiinae, which feed on the alticines, also have acquired the toxicity of their prey and consequently, of the plant itself.

In conclusion, as Fraenkel says, the defensive agent of the plant becomes equally a protection for their insects and at least theoretically, those toxic substances can act as repellents for other insects and predators.

The last problem to be dealt with are the ecdysones and the juvenile hormones, or their analogues, synthetized by plants (phytojuvenil hormones and phytoecdysones). These could act as repellents and may even be transmitted venereally from the male to the female, at least by some Hemiptera, such as *Pyrrhocoris* spp.. In any case, there are elements which prohibit the development of certain insects. It is true that the plants which possess those analogues are more or less resistant and that a "coevolution" has probably developed in that way between the potential host plant and the insects. Insect development is also inhibited by certain deficiencies in essential proteins in the host plant or the increase of tannin content in some.

We return to the phenomenon of larval conditioning, *i.e.*, to the famous Hopkin's principle which was discussed in the previous chapter. That principle, often criticized, however, seems to fit perfectly some known cases among the Lepidoptera. Feeny (1977), quoting recent work, writes that when applying this principle, it seems that adult females prefer to lay eggs on the plant species on which they have fed at the larval stage, and this may help a population to concentrate at a given time on that plant species to which it is most adapted and the one most readily available. Whatever the original contact, consequently, the adaptation is maintained. However, this line of argument does not entirely explain the Hopkin's principle. Can one speak of "phenotypical flexibility" if that term means anything, or of selection among races genetically preadapted? It is very difficult to tell. New research with modern techniques using Lepidoptera might shed some light on the matter. Synthetic media permits modified insect diet and that too needs further investigation. One thing is certain, however, given a choice, species of *Manduca* and *Heliothis* caterpillars always choose the plants on which they originally started to feed, even if it was only 24 hours earlier.

It is to be noted also, as shown in an experiment by Schoonhoven, that conditioning to a synthetic medium can very well greatly modify the trophic preferences of the oligophagous species. The caterpillars of *Manduca sexta* (Sphingidae), a hawk moth, which normally feed only on Solanaceous plants, show, after cultivation on a semisynthetic medium, an ability to eat other plants, such as the dandelion (Compositae), the cabbage (Cruciferae), and the plantain (Plantaginaceae). The caterpillars are generally more adaptable than the beetles, but the normal tendency reappears quickly. Several days of feeding on the normal host plant are enough for *Manduca sexta* to recuperate its old selective habits.

Schoonhoven, in various papers (1968-1972), has summarized the chemosensorial basis of host plant selection by insects, including caterpillars. It is impossible to detail here the wealth of research which has evolved rapidly during the last few years; particularly studies using electrophysiological methods which permit the precise location of the chemoreceptors of the antennae and the mouthparts. An electroantennogram (EAG) response reflects the overall response of the olfactory receptor cell population (Visser, 1982). Among caterpillars, the sense of smell is roughly localized on the antennae and the maxillary palpi. Taste, on the other hand, is situated partially on the maxillary palpi, but mainly on the maxillae, or the epipharynx and the hypopharnyx inside the buccal cavity. Even when two species share the same host plant, the sensory reactions can vary and they can be attracted by different secondary substances, though in most cases the chemical substance involved is the same.

Interesting results have been obtained by removing the maxillae of an insect. Complimentary electrophysiological data show the location of particular sensory areas. The loss of the maxilla permits these insects to accept foods not ordinarily eaten, that is, the loss of food selectivity results. It even permits the consumption of various inert substances. Agar-agar is eaten by *M. sexta,* and filter paper by the Colorado potato beetle and silkworms when their maxillae are excised. When phagostimulants and repellents are combined they produce synergical reactions among insects. More details on this subject may be obtained by referring to Schoonhoven (1968).

As we have noted previously, the difference between monophagy and oligophagy can be quite subtle. Only certain chemical compounds, combined with the absence or presence of some secondary chemical in a plant, could allow such a change. However, a thorough analysis within this chapter is difficult, as it is a phenomenon as complex as the entire topic of food selection among phytophagous insects.

"I care more about *Drosera* than
the origin of all the species
in the World."

-Charles Darwin, 1860

Chapter 6

Carnivorous Plants

Although the general aspects of insect eating plants are not so
astounding, some magazines still report from time to time the old
legend of maneating trees in Malagasy. Several years ago an
issue of a French digest described in detail the agony of a maiden
supposedly eaten by the plant (fig. 3). If truth is stranger than
the imagination of science fiction writers, the actual plants
surpass the tales of these writers. These small and graceful
plants assimilate nitrogenous compounds as a supplement to the
nitrogen poor soils found in peat bogs or swampy water in temperate countries where they grow. Epiphytic plant life in tropical
areas require this dietary supplement as well.

Normally plants produce protein from nitrogen compounds in
the soil. However, plants, as well as animals, have proteolytic
diastase, an enzyme utilized within the cell to change proteins to
amino acids. Within seeds, for instance, this is used for growth.
Some plants can also secrete these proteases which are used for
external digestion. Among them are fungi, bacteria, and insectivorous plants.

Carnivorous plants belong to two branches of dicots, the
Choripetales and the Sympetales (Lentibulariaceae) which have
produced on each branch complex and divergent species. These
carnivorous plants may be divided, according to function, into two
groups: active and passive trappers (fig. 4).

According to some specialists, more than 500 carnivorous
plants exist in the World. These are divided into 15 genera and
distributed in six families of dicotyledons: Sarraceniaceae, Nepenthaceae, Droseraceae, Byblidaceae, Cephalotaceae, and Lentibulariaceae. Two families (see Table 2) are only distributed in
Australia.

FIG. 3. a.- The imaginary tree eating a young maiden in Madagascar. According to legend, the prey is attracted by the smell of nectar and the tentacles surround the captive. Then the folial glands start the digestive process(after "American Weekly," 1920, quoted from Heslop-Harrison). b.- Escaping from the embrace of the alleged "carnivorous trees," the remains of humans digested are seen. This tree is alleged to be in the Philippines, but it is merely a place of human sacrifices ("American Weekly," 1925).

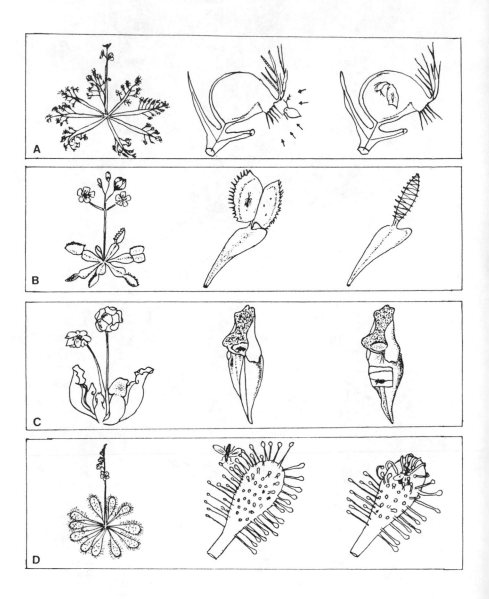

FIG. 4. Active and passive carnivorous plants. a. ⁻ *Utricularia inflata*, an American species; b. ⁻ *Dionaea muscipula*, a species from the Carolinas; c. ⁻ *Saracenia purpurea*, a North American species; d.⁻ *Drosera intermedia*, from the U.S.A.; the genus *Drosera* is widely distributed around the World (after Heslop-Harrison, 1973).

The others are distributed worldwide, chiefly the ubiquitous genus *Drosera* with 90 species which are found in the cold peat bogs in the Northern hemisphere to the tropical mountains of America and Papua New Guinea, in rocky, temporary pools, during the rainy season in the presaharelian zone in Africa, in South Africa, and in Australia. During the dry season, in those areas, the pools dry up. Species of *Utricularia,* aquatic plants that capture prey are also ubiquitous. Finally, to this list might be added certain carnivorous fungi. Some of these, for example, species of *Cordyceps,* invade the bodies of caterpillars, and other insects while others seize nematode worms and other small soil living organisms.

Carnivorous plants probably originated from plants with sticky hairs such as those found on various other plants, for example, some Solanaceae. These hairs have been described as present on several species of *Solanum, Lycopersicum, Nicotiana,* and *Petunia.* This evolution was postulated by Darwin as early as 1875. Among the last two genera, the trapping hairs are both sticky and toxic. A few species of *Solanum,* other than the culti- vated potato, have these trapping hairs on the leaves which help to protect them against numerous leaf eaters. Other plants in the temperate regions, such as species of *Silene* have sticky traplike hairs.

The following hypothesis has been presented, but without a great amount of evidence, that *Dipsacus sylvestris,* a common teasel, with its wrapped leaves which retain rainwater also use small insects as a nitrogen source when they are trapped in these leaf reservoirs. Numerous other plants from temperate regions (these also studied by Darwin) have "precarnivorous" features.

Certain plants, such as species of *Arum* in Europe, hold water in the leaf bracts. This reservoir maintains a small arthropod fauna. In the tropics similar plants are much more important as a habitat for aquatic organisms. The equitan leaves of Musaceae, Bromeliaceae, and similar plants provide a place for the develop- ment of toads and many insects, including mosquitoes. Some of the latter *(Culex* spp.) transmit human filariasis in Africa. In Trinidad malaria is transmitted by *Anopheles* spp. breeding in the leaf bracts of bromeliads. The last is sometimes called brome- liad malaria. The pitchers of some carnivorous plants such as species of *Nepenthes* can also shelter insects which develop normally as in bromeliad or heliconia leaf bracts.

TABLE II

The Main Genera of Carnivorous Plants
(Modified after Lloyd andHeslop-Harrison)

Family	Genus	No. of known Species	Geographical distribution
NEPENTHACEAE	Nepenthes	65	Southeastern Asia to Sri Lanka and Madagascar
SARRACENIAC.	Sarracenia Heliamphora Darlingtonia	9 5 1	Eastern North America Guyana; Venezuela n. California, s.Oregon
CEPHALOTACEAE	Cephalotus	1	Southwestern Australia
BYBLIDACEAE	Byblis	2	Northwestern & Southwestern Australia
DROSERACEAE	Drosera Drosophyllum Dionaea Aldrovandra	90 1 1 5	Ubiquious Southwestern Iberic Peninsula; Morocco Southeastern U. S. A. Southwestern Europe; Africa; India; Northwestern Australia; Japan
LENTIBULARIAC.	Pinguicula Utricularia Biovularia Polypomphalyx Genlisea	30 275 2 2 10	Holarctic Ubiquious Cuba, Eastern South America Southern Australia South America; West Africa
MUSHROOMS	Numerous genera w/traps	20 or more	Ubiquious
TOTAL	15 genera of Dicotyledons	519	

TABLE III
Trapping Systems of the Main Carnivorous Angiosperm Genera
(after Heslop-Harrison)

TRAPPING SYSTEM	GENERA
Active Trap with closing system (snap)	Dionaea, Aldrovandra
Suction trap	Utricularia
Passive Pitcher plant (pitfall)	Nepenthes, Sarracenia, Darlingtonia. Heliamphora, Cephalotus
Adhesive droplets	Pinguicula, Drosera, Drosophyllum
Pseudodigestive tract (suction)	Genlisea

In Malaysia, after a heavy rain, the pitchers of *Nepenthes ampullaria* are filled with rainwater. Mosquito larvae breed there in abundance. Later, when the water has evaporated and the digestive diastasis is concentrated, the mosquito larvae may be killed and digested. However, in almost all cases, carnivorous plants are not simple receptacles, but catch and completely digest small arachnids and insects attracted to them. It is very easy to feed a species of *Dionaea,* the Venus's flytrap, with pieces of meat at home. There is nothing in common between the carnivorous plants and such plants as coniferous trees which trap insects in their resin. This, although taking place since the Tertiary, does not contribute to the food supply of the plant. The resin later becomes amber. This trapping system seems without any clear purpose.

Toads living in the water at the base of the leaves of epiphytic Bromeliacease of the New World tropics are perfectly adapted to this habitat. With enough water for their development, the aquatic insects also living there provide them with food. The toad is perfectly camouflaged and protected by a spiny helmet on the head. Anopheline mosquito larvae which breed there are also specially modified. In contrast, the traptype carnivorous plants do not store water and, therefore, do not have the kind of special fauna found in others of the reservoir (pitcher) type. The latter type possess hairy systems or modified leaves which impale and retain insect invaders.

Sometimes there is an association between pitcher type plants and the carnivorous trap plants. For example, some South American carnivorous Lentibulariaceae live exclusively within the reservoir leaf bracts of the epiphytic Bromeliaceae where they trap aquatic organisms. Such a narrow and unique ecological niche is without equal in any other group except for species of *Utricularia* which are discussed below.

Some species of carnivorous plants live in narrowly restricted ecological niches, such as species of *Heliamphora* (Sarraceniaceae) found in the cloud forest region between Brazil and Guyana and *Darlingtonia californica* (Sarraceniaceae) in swampy regions of Oregon and northern California. Species of *Dionaea* also have an extremely limited distribution in the U.S.A.. Species of *Drosophyllum* from the Southwestern part of the Mediterranean region depend on marine fogs for their water supply. The desert Bromeliaceae of the coastal desert of Peru are unique plants able to survive because of the fogs in an otherwise rainless plain, similar to the *Drosophyllum* cited above.

It must be noted that certain plants, such as species of *Roridula* (Roridulaceae) from South Africa, despite analogies, do not seem to be really carnivorous plants, but rather, when small bugs (Hemiptera) light on the flowers, they trigger the release of pollen. In some others, particularly *Caltha dioneaefolia,* there is still doubt whether, despite their structural similarities to pitcher plants, they are actually carnivorous. Many other "carnivorid" type plants, some of which are, perhaps, carnivorous, are cited by Heslop-Harrison (1976) as belonging to the Dioncophyllaceae, a relict group of great interest because it is apparently related to the carnivorous families Nepenthaceae and Droseriaceae. Among the Scrophulariaceae, which includes *Lathraea* spp., are plants with rhizome type leaves, which are now known to be parasites on other plants and not carnivorous.

Before reviewing the main carnivorous plants, let us note that the pitchers of species of *Nepenthes* sometimes harbor insects which are immune to the digestive enzymes. The plants with stalked glands, such as the sundews *(Drosera* spp.), *Byblis* species, *etc.,* also have a specialized fauna comprised of insects and spiders which suck the juice from the freshly captured insects in the plants. Even a mammal, the tarsier, a relative of monkeys, sometimes visits the pitchers of certain *Nepenthes* (*N. rafflesiana)* in order to collect the captured insects. It cannot, however, penetrate into other species of the same genus because of the presence of barbs in the pitcher. One spider, *Misumenops nepenthicola,* weaves a small web near the attractive glands inside the peristome of some species of *Nepenthes.*

Passive trappers include pitcher plants (species of *Sarracenia)* where the prey is captured and digested inside pitchers or ascidiae, actually modified leaves or the tip of the leaf or the petiole (as in *Nepenthes* spp., fig. 5). There are three theories to confirm the nature of the pitcher of *Nepenthes:* 1) a modification of the petiole; 2) a simple, or composite leaf, or 3) a terminal gland hypertrophied.

Prey is attracted to the pitchers by attractive colors and odors from pseudonectaries at the edge in exactly the same manner as pollinator insects are attracted to flowers. Insects enter the pitcher and drown. These are digested inside the pitcher by enzymes in species of *Nepenthes,* or mostly by bacteria in species of *Sarracenia.*

Another type of passive trap is shown by the species of the genus *Genlisea* found in South America and Africa. These have a "digestive tract" simulating the intestine of an animal. There are underground leaves which have branches in the shape of hollow tubes, each one ending in an utricule or bladder. The tubes are narrowed from place to place where are located spines directed backwards. These intervals are covered with digestive glands. Mud formed from water secreted through the walls of the utricule and the tube flows through the "digestive tract" of the plant. Protozoa, crustacea, and insect larvae contained in this mud form the main food of the plant. The undigested remains accumulate in the utricule. To finish completely the analogy with the animal digestive tract, it would be necessary to replace the running water by peristalic movements and to give an anus to the utricule, but such is not the case.

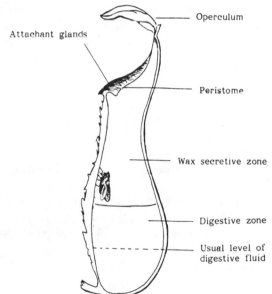

FIG. 5. Longitudinal section through the pitcher of *Nepenthes rufescens* (after Juniper and Burras, 1962).

Among the several active trappers, an American Droseraceae, the Venus's flytrap *(Dionaea muscipula)*, for sale even in shops, is distributed in small areas in southeastern U.S.A.. It grows very well in a pot of sphagnum. This plant has leaves modified into a trap capable of snapping shut when its prey, chiefly small insects, touch one or two of the trigger hairs on the inner leaf surface. The two sides of the leaf snap together, closing the trap. The long spines along the edge of each side of the trap intersect and hold the prey within. A similar closing system is found in *Aldrovandra vesiculosa* (the only species of this Droseraceae genus) found in the Old World. *Dionaea* species have a closing mechanism consisting of a springlike hinge composed of cells on one side with thickened walls. These cells are kept open by their turgidity. The other side is composed of a thin palisade tissue near the central rib. The excitation of the tactile hairs on the ventral surface of the leaf causes the rapid loss of water in the cells. The tension of the thick cell walls brings about the closing of the trap. Although this movement seems animallike, no muscles or nerves are involved. A similar mechanism is found to cause the opening of sori to release the spores of ferns. It is evident that in their natural surroundings these plants are more active than those in cultivation. Only healthy plants in a hot and humid atmosphere react normally. The traps of the Venus's flytrap stay closed for a long time (about 5 to 10 days) if the food is animal matter, but only 5 to 10 hours if the stimulant is inert, mechanical, electrical, or chemical. One may ask: "Why the difference in duration?" Perhaps the trap is kept closed because of the osmotic pressure created by the juices of the captured insects. A salt solution or an extract from another plant which contained an insect will produce the spontaneous closing of the leaves. Re-cently, it has been suggested that the chief stimulus for this action is the uric acid abundant in insect excreta.

Among other active trappers are the unique species of *Utricularia*. There are 150 or more species (certain authors say 275) of these ubiquitous plants in aquatic or subaquatic habitats. Their prey is aspirated by suction cells in the trap set off as soon as the trigger hairs around the mouth of the bladder are touched. The bladder expands suddenly to draw in both water and prey. Then the entrance closes and excess water is excreted by specialized cells in the shape of branched hairs. It takes about 30 minutes to kill the prey, after which the trap is ready to function again.

One utricularian lives inside the reservoirs of an epiphytic Bromeliaceae *(Tillandsia* sp.), a kind of suspended aquarium, retaining water and prey for these carnivorous plants. This utricularian reproduces asexually through runners which grow out of the bromeliad in search of a new host plant.

Other passive trappers have developed fundamentally different systems. There are plants with leaves that are flypaperlike, species of *Pinguicula* and *Drosera* (sundews) for instance. Among these genera we find that the glands on the leaf surface secrete water and adhesive mucilaginous droplets. Insects are attracted by odor, color, and the brilliant refractions of the droplets. When they alight they are trapped in the adhesive substance. When moving to escape from the sundews, for instance, the insect becomes more firmly attached because the tentacules are stimulated to secrete more and more, and the longest, those surrounding the tips of the plant, bend towards the center until they completely cover the insect. After one or two days the tentacules open and reveal only the chitinous skeleton remains.

Generally the prey of these carnivorous plants are small species of arthropods or molluscs, but sometimes even mice have been found inside the pitchers of species of *Nepenthes;* tree frogs in the pitchers of some species of *Sarracenia,* and very small fishes and tadpoles in bladders of some *Utricularia.* The latter normally capture, due to there small size, rotifers, copepods, and mosquito larvae. Species of *Pinguicula* capture and digest atmospheric pollen.

The quantity of insects captured by carnivorous plants is difficult to estimate. The number may be considerable because some pitcher plants are filled with the putrid remains of their victims. Heslop-Harrison, quoting an old text, mentions that a two acre area (8 km) of a species of *Drosera* in England had caught about six million cabbage butterflies with the average of four to seven insects per plant.

Old World species of *Nepenthes* are most common in the forests of Malaysia. Low altitude forest species are few, but there are many endemic mountain species in Sumatra, Malaysia, and Borneo. The color of the pitchers within a species varies greatly and is probably genetic.

Pitcher plants digestive function is possible only by a rather complex process not very clearly seen with the light microscope. The use of an electron microscope, particularly a scanning electron microscope, shows in detail how the trap works. Roughly, *Nepenthes* spp. traps consist of an external stomach, the pitcher, modified, according to some authors, from a leaf or its midrib. The pitcher varies considerably in size according to age and species. When the seeds of species of *Nepenthes* first germinate, the cotyledons enlarge and then the first tiny leaf appears. It is a pitcher which measures about 0.5 cm when first open. Certain large species have pitchers 30 cm to 50 cm long, 16 cm in width and with an opening diameter of 10 cm.

The pitchers of such large species as *Nepenthes rajah* are so large they can capture rats. It has been said even that the very large *Drosera regina*, a South African species with leaves up to 60 cm, can entwine small animals and those species of the Australian genus *Byblis* can trap frogs and lizards. Probably this is a little exaggerated, but certainly carnivorous plants are not exclusively insect eaters, though insects and crustaceans remain 99% of their diet.

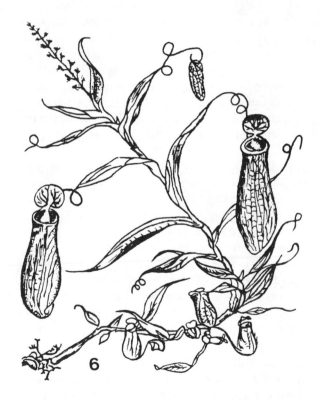

FIG. 6. A *Nepenthes* sp. vine, schematized after several authors.

The early formation of pitchers at the seedling stage in *Nepenthes* spp. (fig. 6) is comparable to the formation of domatia and open holes immediately after germination of seed of epiphytic myrmecophilous Rubiaceae (*Hydrophytum* spp. and *Myrmecodia* spp.). The "mouth" of the pitcher is surrounded by a rim of spicules, the peristome. This is topped by a cover which seals the immature trap until it becomes operational, and later, prevents rain from diluting the digestive juices. Before Darwin's observation it was believed that the "cover" closed after insect captures. Unfortunately for the ultra finalists, this is not true. The sharp pointed barbs provide an obstacle preventing the escape of large insects. As we have seen, it prevents vertebrate predators from stealing the contents of the pitcher.

The pitchers are often brightly colored and have false petals to attract insects. Their odor is also attractive to would be pollinators. Access to the opening is easy for such nonflying insects as ants.

Nepenthes spp. use more than one way to capture prey. Both female and male flowers have sepals covered with small glands producing sweet nectar which attract pollinators. The seeds are distributed by the wind. However, it is the cover and peristome nectar which attract the insects into the pitcher. We can distinguish inside the pitcher three kinds of secretory cells (see fig. 5): a) nectar attracting glands situated under the cover, between the barbs of the peristome, and on the leaf stalk. They are strong attractive to insects; b) digestive glands present in the lower third of the internal surface of the pitcher, and these secrete the "digestive juice" which is essentially a wetting agent used to drown the insect. The actual digestion is apparently caused by bacteria. Also these glands are used to absorb the digested nutrients. Finally, c) the cells of the waxy secretion area extend from just below the peristome nearly to the digestive gland area. With the scanning electron microscope it is possible to study the nature of this relatively complex waxy area. There are really two layers of wax, a basal uniform waxy netting on which the waxy scales lay. It is on these scales that the legs of flies, earwigs, millipedes, woodlice, and thrips slip until they reach the bottom. The same thing happens to snails and slugs even with their sticky integument. Barbs directed toward the bottom prevent insects from attempting to escape by crawling up the side of the pitcher. Certain insects, those with specially adapted tarsi, do not slip, such as a fly which breeds on rotting insects in the pitcher. It can escape as an adult by piercing holes in the pitcher itself. The caterpillars of a moth make a web on the hairs of the pitcher, and a spider, previously noted, weaves a web inside the pitcher. All of these are adapted for walking on slippery surfaces and are immune to the digestive fluid in the pitcher.

Nutrient juices absorbed from the pitchers or traps of carnivorous plants penetrate the leaves with an astonishing swiftness. Amino acids and peptids marked with C 14 can penetrate the leaf in 2 to 3 hours and reach the roots in less than 12 hours. Similar experiments on species of *Drosera* in Australia using Nitrogen 14 show the same results. It must be noted that often among carnivorous plants, the same glands which secrete digestive diastases absorb the results of their digestive action.

Roughly, what characterizes the carnivorous plants? There are traps, mobile or not, color, sweet lures, attractive smells, directional guides, secreting and absorbing glands, and nonreturn systems made of hairs or wax to prevent escape. Really, all of these phenomena are not absolutely unique. *Mimosa pudica* and some Berberidae, show rapid movements of the leaves or the stamens. Some plants, such as species of *Calotropis*, and other Asclepiadaceae, as well as many orchids and Aroideae, have a temporary insect trapping system associated with pollination by species of *Xylocopa*. These plants release their prey once pollination has taken place. This can vary from several minutes to many hours. Numerous other plants have similar features, *e.g.* odor, color, and nectar glands, to expedite pollination. Many other plants, including myrmecophilous species, secrete water, salts, proteins, lipids, mucilage, and sugars from nectaries and pseudonectaries. The uniqueness of carnivorous plants is only that they combine within an autotrophic plant most of these previously listed characteristics.

Thanks to the electron microscope, the use of radioactive marker isotopes, and modern methods of cytochemistry and biochemistry,we were able to clarify some of the still disputed points about carnivorous plants and to confirm the efficiency and finality of their adaptations. Adaptations such as ant guides made by the nectariferous glands on the pitcher of species of *Sarracenia* remain as ingenious as the most perfect adaptations of the entomophilous flowers.

Heslop-Harrison (1978) gives a table of the digestive diastasis secreted by 13 genera of carnivorous plants of the world. He points out that both biologically and cytologically, the secreting and digestive cells of the carnivorous plants are comparable to the animal cells with similar functions (namely the cells of the pancreas). *Pinguicula* species produce secretions of digestive juice sufficiently strong to digest a small captured insect, but a large insect excites an excess of secretion which will run out on the sides of the pitcher when the secretion goes on for several hours. The liquid is so abundant that the digestive cycle is not completed and the nutrients are not absorbed. Then the leaf starts to decay, a case of true botanical indigestion. Similar indigestion is also well known among species of *Utricularia* when the plant, as an animal, bites off more than it can chew!

Before they open, the pitchers of *Nepenthes* spp. are sterile, with a pH of 2.5. After opening, they become rapidly infected with bacteria which aid in the digestion of prey. The pH changes to neutral, and as the result of the decay of prey, a putrid smell is characteristic. Judging from the nature of analyzed diastasis, nitrogen and phosphorous appear to be the elements most useful for carnivorous plant nutrition.

Dictyosomes are particularly active during the secretion of the mucilaginous glands of species of *Drosera* and *Pinguicula*. Not one of the carnivorous plants produces a chitinase, therefore, the exoskeleton of arthropods remain intact after the completion of digestion. There is no doubt, now, that the attraction, capture, and digestion mechanisms which have evolved for insect eating complete for the plants their supply of mineral salts by providing protein nitrogen, and that this allows them to live in habitats where few other plants can survive. Their peculiarity, however, is mostly anatomical, consisting of a trapping system and external secretory glands.

Certainly an evolutionary parallelism exists between the digestion in the pitchers of species of *Nepenthes* and the myrmecophilous pitcher of species of *Dischidia* (Asclepiadaceae), both of which are Southeast Asian plants. The pitcher of *Dischidia* spp. is covered inside with white, adventitious roots, and contains various debris brought by ants which live in it. The rim of the pitcher does not seem to have a precise function. Many pitchers contain rain water, collected humus, and ant excreta. Liquids containing nitrates are absorbed by the internal roots. In some way, two analogous structures (pitchers) have been formed on these two different plants. For both the carnivorous, and the myrmecophilous functions, the evident aim is to add to the quality of the nutrition for underprivileged plants in an epiphytic surrounding.

Contrary to the Latin adage: "*Vulnerant omnes, ultima necat*" (All produce injury, the last one kills), the carnivorous plants do not release their prey and only rarely does an insect succeed in escaping from the trap. It is always rapidly covered with sticky mucilage, and drowned before being digested. The mechanism seems to be almost perfect.

Let us note also that the movements of certain carnivorous plants are not isolated facts in the plant kingdom. It is very possible that the sudden movements of the leaflets of certain *Mimosa* (Leguminosae) and *Biophytum* (Oxalidaceae) efficiently protect the plants against some herbivorous insects which land on them. These movements chase away the insects instead of trapping them.

CHAPTER 7

Myrmecophilous Plants

As Guérin, the French romatic writer explains at the head of this chapter, "Kings can see their palaces falling down, but ants will always keep their dwellings intact." At the beginning of the evolution of myrmecophilous plants, *i.e.*, "ant loving plants," there was certainly a phenomenon similar to the one seen among numerous plants having extrafloral nectaries. Some edible species of *Hibiscus*, such as okra (*H. esculentus*), and many other plants from the temperate and tropical zones have sweet extra floral glands which often attract extremely aggressive ants. A similar phenomenon is known among ferns where sweet secreting glands, true nectaries, are not rare. Plant protection by ants has been called "set a thief to catch a thief," *i.e.*, robber ants (nectar eaters) protect the plant from other possible thieves, including other species of ants. It seems probable that many of these plants, even though not accommodating ants, are protected from herbivorous attacks by the same ants that visit their false nectaries. This is the beginning of a symbiosis which foreshadows a system infinitely more complex, the relationship found in true myrmecophilous plants.

It is evident that some ants colonize dead trees and chew into the wood, but they do not eat this material as termites do. This, of course, is different from the true myrmecophily.

Myrmecophilous plants are unequally distributed in the tropical world. American and Asian rain forests are the richest in such adaptations, while the African zone is the poorest and the most monotonous, lacking any sensational innovations. Nonetheless, it remains true that innumerable plants in the New and Old World Tropics are, in some manner or another, myrmecophilous, though not necessarily displaying the level of adaptational sophistication achieved by several species of neotropical *Acacia*.

FIG. 7. An aerial nest made by ants.Note the ant garden with young shoots seeded by the ants (after Ule, 1906, modified).

Since the first simple ant associations, that of the tree dwelling ants living within the cavities existing in trees took place, a great host of modifications has evolved. Species of *Oecophylla* form a nest by sewing together leaves of trees with silk secreted by their larvae. Far more complex associations of the type found on species of *Myrmecodia* and *Cecropia* trees exist. Ants gather extrafloral nectar, whether they live in cavities they have excavated, or live within true "myrmecodomatia" or preexisting dwellings, they always protect the plants they live on from insect pests. Many of these associations are very complex. In addition to prefabricated housing, the plant may provide glucids, protids, and lipids as well. Arboricolous ants, outside of the special adaptations which will be discussed later, have devised ingenious associations with epiphytic plants living on trees in tropical forests. It is mostly in Tropical America that these associations have reached maximum perfection. Orchids and bromeliads provide ideal sites for building a nest. The epiphytes use the ant excreta as their source of nitrogen, a kind of suspended soil. All species of American ants of the genus *Azteca* and several other genera live exclusively in paper nests on trees, in cavities, or in various ways, under or inside leaves and trunks. The narrow heads and elongated shape of the queens and workers of many of these species evidently are adapted to life inside small cavities. At least one species of *Azteca* is green, an exceptional color among ants. This is probably an example of homochromy. Some soldiers of species of *Colobopsis* close the entrance of their galleries with their truncated heads. Many ants, such as species of *Oecophylla* in the Old World, fight aggressively when species of the arboricolous cicindelid genus *Tricondyla* (Coleoptera) living in Malaysia defend themselves by playing hide and seek around the trunk. However, as we have often seen in the Solomons, ants don't hide—they bite!

In Amazonia, certain ants produce what have been called "ant gardens" on branches of trees (fig. 7). These suspended gardens are started much like ant hills. When ants bring in oil seeds (elaiosome) which they use as food, some of the seeds germinate and their roots bind together to form the garden. These gardens, made at random, however, are advantageous to the oil seed plants as well as to the insect.

Species of tropical ants of the genus *Oecophylla*, as well as many other animals such as Galapagos finches, and the wasp *Ammophila urnaria*, previously described by the Peckhams, use living or dead "tools." Although using manufactured tools is unique to humans, many animals use natural objects as tools. For example, these ants squeeze larvae, producing silk used to stick together the edges of leaves which make a part or all of this nest. The workers are lined up to make a living chain to bring together the sides of the leaf (fig. 8). The queen of *Oecophylla smaragdina*, the most common species in Southeast Asia, is green, allowing her to establish her nest on an opened leaf without attracting too much attention from predators. These ants, strongly aggressive, protect fruit trees from phytophagous beetles, a means of control in use in China. Even if the ants raise aphids and coccids, they seem to be more useful than harmful. Controlling them with insecticides, as is done in some countries, is certainly a serious mistake, rupturing a fragile biological equilibrum.

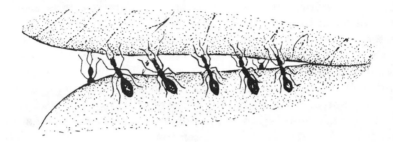

FIG. 8. Species of *Oecophylla* ants bringing together leaves which will later be "sewn" together with the silk produced by their larvae and used to establish a nest (after several authors).

Let us come back to the true myrmecophilous plants, which have been called "myrmecophytes." They have been classified as follows: myrmecotrophic (providing food), myrmecodomic (offering a dwelling [fig. 9]) and myrmecoxenic (offering both together). In all, a lexicon of terms has been created, such as the term myrmecodomatia or domatia ant dwellings, which can be cauline (in the stem [fig. 10]), or foliar (in the leaf), or in various other sites such as in the calyx, spines, or stipules.

In 1875, Darwin wrote a classical book on the carnivorous plants, a similar book still remains to be written on myrmecophilous adaptations. The literature on the latter is scattered in a great number of papers, essentially devoted to American, African, and Indo-Australian floras. The phenomenon is practically unknown in the temperate zone, with the exception of the extrafloral nectaries which are world wide in distribution as in certain species of *Prunus*.

Among the large number of described myrmecophytes, we choose to describe here only the following few examples. Myrmecophilous relationships are known among the vascular cryptogams (some ferns), the monocotyledons (various palms and orchids), and many dicotyledons (species of Asclepiadaceae, Boraginaceae, Bromeliaceae, Euphorbiaceae, Flacourtiaceae, Gesneriaceae, Monimiaceae, Moraceae, Meliaceae, Myristicaceae, Passifloraceae, Piperaceae, Polygonaceae, Potaliaceae, Rubiaceae, Saurauiaceae, Sterculiaceae, Scrophulariaceae, Verbenaceae, and some others). In all, more than 25 families are represented by more than 41 genera and 270 species. About 15.5% of the species are of African origin, 43% American, and 41.5% Indo-Malaysian. The morphology and evolution of the myrmecophilous structures in plants have been studied in detail by Schnell (1970), and we will discuss several cases representing the most typical from forests in both the New and Old World. I have personally (Jolivet, 1973) reviewed succinctly some very typical cases in the Malaysian Region.

FIG. 9. A nest of leaves sewn the with silken secretions of the larvae of *Oecophrlla smaragdina* (after Frisch, 1974 and Dumpert, 1978).

The American flora exhibits the most remarkable adaptations. Indeed, one of the most recent authors on the subject, Janzen (1966), shows that there is an effective symbiosis between species of ants of the genus *Pseudomyrmex* and several species of *Acacia*. For instance, the mutual benefit realized by *Pseudomyrmex ferruginea* and *Acacia corniger* is evident. The ants within the hollow enlarged spines of the tree (fig. 11), feed on the sweet secretions of the four nectaries at the end of each leaf stalk, and protect the trees from herbivorous vertebrates and invertebrates. The ants also cut off the modified extremities of certain leaflets of *Acacia cornigera* (belt or beltian bodies) to feed their larvae. These leaflets are rich in proteins and lipids. It seems that one of the main goals of the ants is to relieve trees of vines and invading plants by chewing and destroying the young shoots of these invaders. The efficiency of this method is apparent after the ants disappear, for one reason or another. Phytophagous insects attack the foliage of the *Acacia*, while the surrounding plants cover the tree, causing it to die within three to twelve months. Other *Acacias* survive without ants but they have different chemical properties. It has been shown by Rehr, *et al.* (1973), for instance, that in Central America, the species of *Acacia* protected by ants is perfectly edible to the caterpillars of *Prodenia eridania* (Noctuidae), whereas the species of *Acacia* without associated ants are toxic to many herbivorous insects. Apparently, the quantity of cyanogenetic heterosides is responsible for this difference. Experiments using artificial media, made from plant extracts, indicate that this is true. Also to be noted is that other myrmecophilous plants differ from their ant-free relatives in their lack of toxic latex. The latter directly provides the same protection offered by ants.

Ant-plant association in one species of *Acacia* begins when the young shoots are colonized by species of *Pseudomyrmex* after the plant reaches a size of about 25 cm. Even though they have only one stem, several leaves, and a pair of spines, the spines are rapidly invaded by the ants. An opening is made on the lower side of the leaf where it is protected from rain. Also, it seems that there is here a specialization of tasks among the workers. Some collect nectar or beltian bodies, others tend the larvae.

Another American plant, a species of *Cecropia* (Moraceae [fig. 12]) is "inhabited" at the stem internodes by ants of the genus *Azteca*. Food is provided by the plant in the form of mullerian bodies (trichilia) located at the base of the leaf stalk among dense hairs. In this case, sugar is not provided by the plant, instead, the ants rear some scale insects (Pseudococcidae) inside the internodes. By obtaining honeydew from the scale insects, the ants are able to compensate for that deficit.

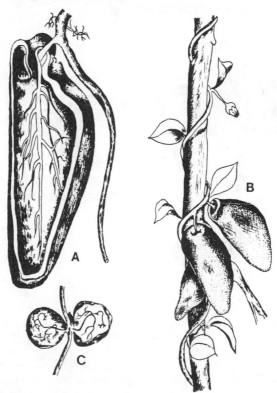

FIG. 10. Myrmecophilous plants from Southeast Asia. a. *Dischidia rafflesia-na* (Asclepiadaceae). Section through a big leaf showing the internal roots of the morphologically inferior but internal face. b. Plant with its two kinds of leaves, the samm and short ones and the pitcher leaves with ants. c. *Dischidia* sp. with small round leaves, the inferior face showing the roots (after Holtum, 1954).

Janzen (1969) has proven that, as in species of *Acacia*, *Azteca* ants clear away the fast growing shoots of vines infesting the cecropia trees. These vines climb the trunk of the plant and would soon cover the plant if not removed. This "allelopathy" of the ants is functionally similar to the chemicals which other nonmyrme-cophilous plants produce in order to defend themselves against plant and animal invaders, mostly leafcutting ants of the genus *Atta*. In place of the chemicals, the plant, at high metabolic cost, must feed and shelter these ants. According to Janzen, the ants constitute an allelopathic agent more efficient than the modification of the plants chemistry. In both cases, there is a deficiency in the energy budget of the plant.

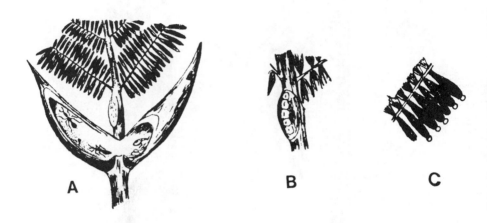

FIG. 11. A Neotropical *Acacia* with: a. Hollow spines which shelter the ants. b. Nectar bodies on the rachis. c.Belt bodies (proteins) at the top of the folicles (after many writers).

To be noted, however, cecropia trees tolerate predators which cause some damage, such as sloths *(Bradypus tridactylus)*, several caterpillars, and leaf beetle larvae. The same thing happens with almost all the myrmecophilous plants that we have observed. Several caterpillars (including the symbiotic lycaenid butterflies), and some weevils (such as species of *Oribius)* are tolerated in New Guinea by ants of the genus *Iridomyrmex,* small but ferocious guards on species of *Myrmecodia* and *Hydnophytum* (both Rubiaceae). The American flora is so abundant that it would be impossible to list here all the genera of Neotropical myrmecophytes.

In Africa, however, the specialization is less developed. The East African *Acacia*, the remarkable adaptations of the American species absent, have hollow stipules inhabited by ants. These are natural structures, not induced by the ants which appear irregularly on the plants. Also, Janzen mentions that in Africa the myrmecophilous species of *Barteria* (Passifloraceae) are efficiently protected by the ants, first against the vertebrates (such as *Colobus* sp., an Old World monkey), and invertebrate predators, as well as against adventive and parasitic climbing plants. The species of *Barteria* "occupied" by ants have more leaves, more branches, and a healthier appearance than the antless species.

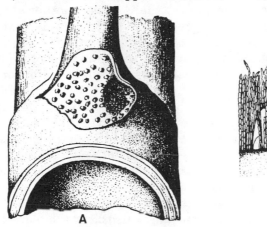

FIG. 12. *Cecropia adenopus* (Urticaceae), a tropical American myrmecophilous plant. a. Base of the leaf stalk with pad, and b. Muller bodies at various stages (after Gadeceau, 1907).

In Asia, the adaptations are very sophisticated and, in some points of view, are somewhat similar to the New World plants. In my paper (Jolivet, 1973) on the myrmecophilous plants of Southeastern Asia, I have described in detail the extraordinary adaptations of species of *Myrmecodia* and *Hydnophytum* (Rubiaceae) with ants of the genus *Iridomyrmex*. The relationships found are diverse. Often, as in the species of the genus *Dischidia*, the leaves offer dwellings for the ants. Sometimes these are the foliar sheaths, as among the rattans *(Korthalsia* spp.); sometimes the hollow internodes of the stems, and the stalks of these plants are inhabited. In some other cases, the tubers of epiphytic Rubiaceae are inhabited by ants or even the rhizomes of certain tropical ferns (species of *Polypodium* and *Lecanopteris*). The myrmecophilous plant species are vines, epiphytes, and such forest trees as species of *Macaranga*. These are found everywhere except in the alpine zone. Just below that zone, however, epiphytic species of the genus *Myrmecodia* (Rubiaceae) retain their myrmecophilous relationship but become ground plants growing on mosses, including sphagnum, an environment as poor in nitrogen as that of their arboricolous surroundings. In New Guinea, this subalpine zone corresponds to the arborescent fern zone and the northern limits of the forest which is reminiscent of New Zealand scenery. Interestingly enough, the species of *Dischidia* already mentioned do not always have "their" ants. Another ant, much more aggressive and not living in the plant's pitcher but on the trunk, may prevent the normal inhabitants from settling on their host. This ant collects waterproofing wax from inside the pitchers, and then the internal roots normally absorb the water contained in the reservoir.

It would be too time consuming to summarize the multiple adaptations shown by Asiatic myrmecophytes. As for the American species, the ants efficiently defend the plant against most vertebrate and invertebrate predators, although there are other predators, including moths and beetles, adapted to the plants. Also, a small fauna of worms and mites are commensals inside the plant, as well as some staphylinid beetles which are, as yet, undefined. As in other regions, ants also clear the plant of vines and other rival plants.

Many plants, particularly epiphytic Rubiaceae, develop internal cavities before they are invaded by ants. Internal warts found in species of *Hydnophytum* (fig. 13) and *Myrmecodia* absorb nutritive nitrogenous substances formed by excreta, detritus, and dead ants. In exchange, the plant provides sugar in the form of colored and sticky berries similar to the false berries of mistletoe. As in the epiphytic Loranthaceae, the berries are either red or white according to the species. It does not seem that the aposomatic color of fruits are attractive to birds, besides, it would be difficult for a bird to land on an ant nest. Ants, apparently, are responsible for the scattering of seeds, a process termed myrmecochory. Postfloral nectar, *i.e.*, the nectar produced after blossoming and pollination, also feeds ants which live inside the plants.

It is not difficult to see that ants play an important role for these plants: defenders and scavengers, nutrition providers, seed disseminators, and gall generators. In exchange, the plant provides shelter and often food.

Let us conclude with the remarks made (Jolivet, 1973) about the myrmecophytic flora of Southeastern Asia:

1) The inflated and hollow tubercles, pitchered or modified leaves, inflated and modified internodes, are, as are the false galls, or myrmecodomatia, of species of *Acacia*, hereditary and fixed structures.These occur without any (or only slight) intervention from ants, probably the result of a long selection process. The phenomenon is spontaneous and not induced.

FIG. 13. Another myrmecophilous plant from Southeast Asia: *Hydnophytum* sp. (Rubiaceae), showing the inflated stem of the plant with external roots and the holes through which the ants come and live.Flowers and fruits are produced generally at the leaf axis (after Holtum, 1954).

2) The diversification of the Indo-Malaysian myrmecophytes is great, but evolution does not reach the perfection here that is found in certain American species with false nectaries, protein and lipid producing glands, and beltian and mullerian bodies. Myrmecotrophy exists, but often in a more discrete form such as the white bodies of the Macaranga, white and red edible fruits, floral, and extrafloral nectar.

3) Myrmecophily is not exclusively linked, as some may think, to mangroves in Asia. It can be found in abundance everywhere in

and about the forest. It is true, however, that the mangrove areas and the epiphytic and subalpine areas have something in common: a poverty of the soil which is partly compensated for by myrmecophilic symbiosis.

4) Whatever the explanation given for myrmecophilism, whether it comes from Darwinian selection of advantageous characteristics or from the Lamarckism (fixation of induced features, but how can people be Lamarckists in the 20th Century?), the cecidian or galligenous theory is still partly acceptable. It is true that in certain cases the presence of the ants encourages the ultimate development of the plant, even if it does not start it. *Hydnophytum* species inside greenhouses, without their ants, are hollow, but small and undeveloped.

5) The symbiotic theory, strongly opposed 50 years ago, at a time when it was the fashion to deny any finality in nature, still has many supporters and seems to be of value only to the field naturalist who has been able to observe plants outside of an herbarium. This theory cannot be generalized for all cases. It remains, however, that very often the ants provide defense against predators as well as nitrogenous substances, and the plant provides shelter and food.

6) Ants come to preexisting structures with an opening, as in Rubiaceae, or, in other cases, without a visible opening, but with places easy to pierce, as in species of *Vitex*. In either case, there is preadaptation.

7) It is absurd to pretend that myrmecophily is a parasitism harmful to the myrmecophyte, because without their ants, many of those plants wither and die or fall victim to herbivorous animals or vines. Without ants, epiphytic Rubiaceae grow very little, probably for lack of nitrogen, but also, for lack of galligenous stimulation, whatever that may be. Other plants in similar situations suffer from invading vines. The myrmecophilous function seems to be a reality. Let us at least say that in many cases, the ants are useful to the plant and that in other cases, their presence may be indifferent. The term symbiosis is not too precise and we cannot deny what is evident.

8) In response to the debate about host specificity of ant plants in Indo-Malaysia, it is very difficult to formulate a principle. The species of *Iridomyrmex* or *Myrmecodia* are host specific, but those of *Hydnophytum* are found on a few species of plants. Several cases of cohabitation of sympatric species of ants are known. Other myrmecophytes are colonized by different species

according to the area and certain species of ants inhabit several myrmecophytes representing different families. Although not a precise rule, some groups of ants have clear tendencies to myrmecophily and cannot live except on those plants.

Finally, because of our fragmentary knowledge, a general study of the myrmecophyte phenomenon is necessary and considering what is now known, it seems clear that the symbiotic theory will be strengthened by more observations.

In addition to the ant domatia, there are plants with noninsect domatia, structures developed by the plant which gives shelter to other small arthropods including mites (in acarodomatia). To speak of an acarophilic relationship is rather premature, but many of these structures seem preadapted and hereditary. Acarodomatia should not be confused with galls which are induced by some animals. These acarodomatia often provide excellent taxonomic criteria for the separation of species of mites. Some works report that the mites clean the plants or the leaf and thereby remove several harmful elements. Schnell (1970) gives a detailed study of that still debated subject. These domatia, very distinct from the acari galls, are theoretically useful, with little doubt that this actually occurs.

The nonmyrmecophilous plants, for example, the tree *Goniothalamus ridleyi* (Annonaceae) from Indo-Malaysia, has a very strange relationship with ants. This tree is cauliflorous, *i.e.*, it produces its flowers at the base of the trunk. There small ants accumulate soil and bury the flowers still in the bud stage. This is an example of artificial geocarpy. These ants then feed on the nectar of the flowers. The buds of the buried flowers are not accessible to any other insect and hence the ants are the exclusive pollinators. Other flowers of the same plant, not covered, attract many other insects.

Obviously not all plants are myrmecophilous. Some might even be called myrmecophobic because they actually destroy ant nests, for example, certain fungi, mosses *(Polytricum* spp.), and sphagnum. Other plants protect their nectaries from ants, with hairy barriers, preserving them for true pollinators, such as bees. Many plants, for example, the sunflowers and coniferous trees, entangle the legs of ants in their sticky secretions. Sometimes this can be a legitimate means of defense for the tree, but, if true, it is at a considerable energy expense through the loss of latex or resin.

Southwood (1973) has classified the relationships between

insects and plants into eight groups: 1) plants giving food to the insect; 2) shelter; 3) transportation; 4) aids to reproduction, and 5, 6, 7, 8) those where the insect gives the reciprocity to the plant. Examples of these are numerous as we will see later in the following chapters.

We have seen, in the study of myrmecophilous plants, that these relationships are infinitely more complex and that reciprocal services provided are numerous and diverse.

TABLE IV

Relative Importance of the Interactions of insects & Plants
(after Southwood, 1972 and Hocking, 1975)

	Shelter	Food	Transportation	Reproductive Aid
Insects provide plants with	+*	+	++	+++
Plants provide insects with	++	+++	+	+

* += rather common; ++= common; +++= very common

In general, plants exhibit an extraordinary array of adaptations to their environment and myrmecophily is only one case among many others. A. and M. Larson write: "Man in fact is trying to reach the stars. Perhaps, he will reach them one day, but life is such that if he succeeds, it is probable that ants will go there with him." So far they have followed him in trains, boats, and planes. Why not spaceships?

A short sentence from Darwin might be our conclusion: "The brain of an ant is one of the most marvelous atoms of matter of the world, perhaps more than the brain of man." The father of the evolutionary theory was perhaps a little too anthropomorphic at times, but it is a fact that ants are very successful, are everywhere, and have adapted themselves to all kinds of climate and conditions.

"What do I care for tulips and roses since,
By mercy of Heaven,
I have for me alone
the whole garden."

-Hafiz, Les Chazels.

CHAPTER 8

Epizoic Symbiosis

Symbiotic associations between plants and insects have developed in ways other than as myrmecophilous plants. In the mountains of Papua New Guinea, the late J. L. Gressit described a phenomenon of close association between cryptogams and the cuticle of beetles which he termed epizoic symbiosis.

These cryptogamous gardens have been discovered only on big weevils with fused elytra, belonging to the subfamilies Leptopiinae, Brachyderinae, Cryptorhynchinae, Otiorrhynchinae, and Baridinae, and also on a species of *Drytops* (Colydiidae). It is probable that these associations have developed on other beetles yet to be discovered.

Parasitism by algae and fungi are known to occur in human and animal tissues. Algal infections are really rare in man, although certain fungi are highly pathogenous to man, being saprophytic on skin and mucous membranes, often in competitive equilibrium with the bacterial flora. Such is the case of athlete's foot, of tinea caused by *Candida albicans*, and many others. Other cryptogams probably have a neutral or somewhat symbiotic role, as for example, the South American sloths with their associated blue algae. It is also known that scabs on the shell of aquatic turtles are invaded by blue or green algae, filamentous or encrusted, and that giant Galapagos tortoises carry lichens on their shell. On the female turtle, these lichens are eventually worn off due to friction during copulation. Many fish also have algae infested gills, the algae living where there is an optimum amount of oxygen. These are only a few examples of many similar relationships.

The association of algae with aquatic invertebrates, including insects, is very common. Sometimes it is a true symbiosis.

Examples of this are the marine turbellarian, *Convoluta roscof-fensis*, associated with the alga *Phytomonas convolutae;* the giant marine mollusks of the genus *Tridacna* with algae on their gills; the gastropod, *Elysia atrovirdis*, which crawls on leaves and uses free chloroplasts from marine algae; several fresh water hydra with chloroplasts in their bodies; marine anemones *(Anthopleira* spp.), and some sponges, each with algal associations. Unicellular organisms using algae symbiosis are represented by chlo-rophyllate rhizopod and flagellate protozoa. It is because of this, their position in the animal or plant kingdom that is still debated.

What advantages do sloths get from their algae which are so well adapted to the hairs of fur? Perhaps a sort of camouflage; perhaps nothing. However, marine *Paguras* spp. (hermit crabs which inhabit a discarded mollusk shell) seem to be well protect-ed by the garden of algae planted over their adopted shell. The terrestrial *Paguras* (cenobites) and *Birgus latro* (coconut crabs) have lost that kind of protection and the latter have abandoned their shells. The cenobites spend their days like woodlice under stones. Species of *Birgus* live in holes like rats.

Eccrinales, fungi that live in the digestive tract of insects, are a kind of nonpathogenic fungus. No algae are known on insects if we exclude some depigmented Schizophyta nearer bacteria than algae. There are algaellike organisms *(Oscillospira* spp.) in the intestines of rodents and others.

External insect-plant relationships do really exist, less for algae than for fungi, but at least rarely in the first. Among the fungi there are Laboulbeniales and several other groups which are not really pathogens or ectoparasites on the insect cuticle. Green algae have been found rarely on the wings of grasshoppers, on a Chinese geometrid moth in high forests, and on a Queens-land spider. This is exceptional among terrestrial arthropods, but very common among aquatic species such as mosquito larvae, larvae of Chironomidae, and dragonfly naiads.

Gressitt describes (fig. 14) the most typical and complex associations between Coleoptera and cellular and vascular cryp-togams in the high mountains of Papua New Guinea (Gressitt, 1965-1977). The vegetation that covers the cuticle of species of large weevils of the genus *Gymnopholus* is reminiscent of the algal flora which adorn pagurid crab shells in the sea. Probably the weevil flora functions as a camouflage against possible preda-tors at that altitude. At first, Gressitt thought that birds of para-dise were possible predators, but as these birds are becoming increasingly more rare, they do not often feed on these beetles. The most likely predators are marsupials such as species of *Antechinus, Petaurus,* and *Eudromicia.* The weevils do not seem to be protected by repulsive toxic blood, as are many chrysome-lids.

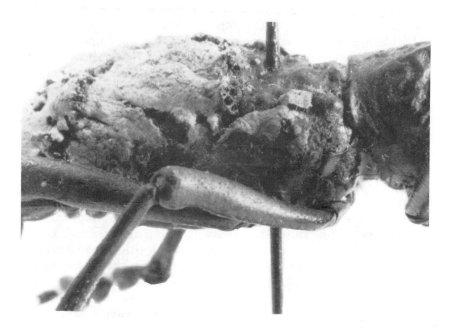

FIG. 14. Several species of beetles exhibit epizoic symbiosis. The abdomen of *Gymnopholus lichenifer* Gressitt, a weevil from New Guinea is picture here (Courtesy of Florida Department of Agriculture, and the Bishop Museum).

Gressit coined the term "epizoic symbiosis" for this new form of external association between a plant and an insect. Oddly enough, this "symbiosis" does not seem to exist (or has not yet been discovered) in the high mountains of tropical South America, Borneo, or even in the Indo-Malaysian area, where local conditions very similar to the cloud or moss forest of Papua New Guinea sometimes exist. Epizoic symbiosis is completely lacking in tropical Africa, where the high mountains are either volcanous or old crystalline mountain chains (*e.g.*, Mt. Ruwenzori). It is rare to see large insects (such as galerucine beetles) above 3000 m, an altitude, however, which would be sufficient to produce the phenomenon. One must also understand that even if those associations were visible on the living insect, one would have to be a specialist to recognize it. On dry specimens, in collections, the vegetation dies and is practically invisible.

In summary, this phenomenon, concerning only weevils and a few other beetles, seems to be extremely rare. Indeed, it is possibly a unique occurrence in the Papuan New Guinea subcontinent, without parallel elsewhere, a fact which is really surprising.

This epizoic symbiosis provides certain Coleoptera with a living garden, mostly on their back, and even more intriguing, a suspended garden inhabited by many small animals. This ecological association includes various cryptogams such as green and blue algae, fungi, liverworts, lichens, mosses, and even in certain cases, the gametophytes of a fern, an oribatid mite, rotifers, nematodes, and various microorganisms. This phenomenon is limited to the "mossforest" altitude, a zone of fog and perpetual rain, as we have seen, which exists at an altitude ranging between 1200 and 3600 m. From 1965 to 1977, Gressitt and his collaborators made many observations on this association and progressively their theory has evolved. I personally visited Gressitt in 1969, at the Wau Ecology Research Station in eastern Papua New Guinea in the Kaindi Mountains where I conducted many experiments. In the gut of a weevil, I found a gregarine related to one from an African weevil. However, during the research in central Papua New Guinea, I have rarely met these weevils. Either they are rare, or they were very well camouflaged. Gressitt gave many reasons for the scarcity of the species, but predation seems to be rare since the duration of their life is long, five or more years. Not only is considerable time necessary for establishing the flora, but old specimens may eventually lose their plant cover.

Can we really use the word symbiosis in this complex association? Yes, if there is a real camouflage for an apterous and defenseless insect, one very slow moving, and already living in the shade of a forest of cryptogams. However, the debate continues, even though there is no question that the association exists and it is very remarkable.

Lichens live on species of *Dryptops* beetles (Colydidae), but it is on species of *Gymnopholus and Pantorhytes* weevils that are found the most complete series of plants. The cryptogams are numerous and varied, sometimes ascomycetes with red, orange, and black bodies, some Fungi Imperfecti, various Cyanophyceae,

green algae, lichens of the families Physciaceae and Parmelia-
ceae, liverworts, mosses, and even, as we have seen before, fern
prothaliia. In the middle of that flora, and probably on it, we find
many oribatid mites (species of *Symbioribates*). These acarina are
very small (+0.2 mm) and are completely hidden beneath the
fungus cover. The movements and mating activity of the weevils
cause fungal spores to be disseminated. Wind must also play its
part in this dissemination, and the adhesive secretions of the
elytra of other weevils collect them by chance. Along with the
fungi also live rotifers, nematods, diatoms, and other microorganisms, all very small. Occasionally, some Psocoptera are seen feeding on the algae and fungi. These host beetles are also parasitized
by some mites, often phoretic, and having nothing to do with the
previous complex association. These mites stay on the ventral
surface, antennae, and legs of the beetles.

The species of *Gymnopholus* are found between 900 and 3600
m and the high humidity of these areas is evidently conducive to
the development of the original flora. It is, however, above 2000 m
that most of the *Gymnopholus* species can be found, or at least
the ones with abundant flora on their backs.

A mountain species of *Gymnopholus*, subgenus *Symbiopholus*,
shows almost exclusively these associations. The other species of
the genus usually have smooth elytra and live at low altitudes.

The morphological adaptations of *Symbiopholus* spp. seem to
encourage the development of the plants. Their elytra have big
hollows, holes, crevices, tubercules, edges, special scales, setae,
and setal tufts which seem to provide places for retaining the
flora. Sticky secretions in the hollows seem to be there to catch
the spores and to encourage the growth of algae.

These insects are sedentary, apterous, with fused elytra. They
live a long time and are relatively polyphagous. As are many
forest Curculionidae in the tropics, they are found on a great
number of bushes and small trees belonging to the same ecological habitats (16 families in all). The larvae are endogeous and
feed on roots.

The plants fix themselves on structural modifications of the
elytra, thorax, and sometimes on the femora. Often the flora
coincides with the hairiness of the beetle. A waxy secretion at the
posterior edge of the pronotum can also help to affix the spores.
Young beetles have their scales intact and are without flora,
while mature specimens are covered with lichens. Nonetheless, as
with the Galapagos tortoises, as a result of rubbing and mating,

the old specimens lose their plants.

These beetles appear to be still evolving. Each part of the mountain chain has its own species. As with the species of the related genus *Cratopus*, in the Mascarene Islands, in the Indian Ocean, the species of the genus *Gymnopholus* seem to be in rapid evolutionary change.

It is important to note that the cryptogams found on the weevils are found also on the leaves and trunks of trees in humid moss forests. The hypothesis, that their protection is the same as that of the shell used by pagurid crabs, is attractive, and indeed, it may be true. If not, let us say that this association is remarkable even if it is only the colonization of a preadapted chitinous structure in a favorable humid surrounding.

In conclusion, at the present time, epizoic symbiosis of insects has been identified nowhere else in the terrestrial animal world. There is only a rough analogy between it and the blue green algal associations in the hair of the Neotropical sloths or the pyralid caterpillars and moths which are also established there. However, the Lepidoptera seem to feed only on the desquamated skin and cutaneous secretions of the sloth, as do the Mallophagan lice. If they were feeding on algae, the association would be more complex. Anyway, it has been proven that the caterpillars once lived only on the excreta of the sloths, on which the moth imagoes still lay eggs during the sporadic defecation of the host. More than a camouflage, the algal cover of these mammals appear, as in the Papua New Guinean weevils, to be the result of the permanent humidity of the rain forest and the quasi-immobility of the beast.

Recently, in Ohio, Slocom and Lawrey (1976) discovered that the larvae of the neuropteran *Nodina pavida* (Chrysopidae), were carrying an accumulation of cryptogamic vegetation on their backs, a sort of camouflage against predators. This is composed of lichen soredia and thalli, pieces of bark, parts of moss gametophytes, pollen grains, fungus spores, and other debris from plants and insects. The phenomenon of an epizoic lichenic flora may not be localized in Papua New Guinea exclusively, and may be much more widespread. Further, this too is a case of symbiosis because the lichens are dispersed by the insect.

Related examples deal not with living plants but the trash that camouflage the nymphs of Reduviidae, larvae of certain chrysomelid beetles, and others. Most of the trash is composed of dead insects, feces, caste skins, and debris, and usually does not involve living plants.

"A hen is the only way
an egg can make another egg."

-Samuel Butler

CHAPTER 9

The Galls

Many definitions of a gall and gall formation have been published, but their origin and production still remains rather mysterious despite the biological and taxonomic work which has been performed on them. Many new hypotheses have been offered, including a proposed change in genetic information supplied to the cells by the gall inducers. No entirely satisfactory explanation accounts for gall formation, and nothing will be proven until one has succeeded in creating a completely artificial gall on a plant.

The dictionary gives the following definition of the word gall: "an external tumor produced by a plant, being its reaction to a small wound containing an irritating agent, in particular that of an insect ovipositor, which deposits one or several eggs in the plant tissue." Another term for this phenomenon is cecidia. Galls caused by mites are acarocecidia, by nematodes are nematocecidia, by fungi are mycocecidia, and by bacteria are bacteriocecidia. Insect galls, which are the most common of all, are called entomocecidia.

Oak galls produced by cynipids (Hymenoptera) are the best known and, because they are rich in tannin, the most exploited. Other cecidia produce tannin in tamarisk in Morocco (the acarocecidia) and in sumac (the entomocecidia).

Galls develop on any part of the plant: leaf, stem, root, fruit, or flower. They are extremely variable in size, shape, and color. Some are very beautiful. Some galls exist on cryptogams including lichens, and on various phanerogams, mostly on dicotyledons. Insects and mites produce 80% of known cases of arthropod galls.

Galls were once considered as harmful to the plants, but now the tendency is to consider the gall as advantageous since it protects the plant against possibly even greater damage, the complete necrosis of the parasitized tissues. Indeed, it seems better for the plant to encapsulate the parasite and to provide it with food than to have the parasite moving around and destroying tissue. It is quite similar to the encapsulation of certain nematodes by animal tissues. The formation of a gall, which is an hypertrophy or dysplasy of a tissue, restricts the damage not only to a certain region of the plant, but also to a particular time in the plant life cycle. It is a defensive and protective measure which seems to be the best way to combat this invasion. A gall has been likened to a false fruit, but since this has never been totally reproduced in the laboratory, the analogy is not certain.

Before discussing the various types of galls, one should mention that certain species of small ants invade the oak galls after they have been evacuated by their prior inhabitants in the oak tree. As the colony grows the ants also leave and migrate to a larger habitat. Also, when considering ant and gall relationships, the case of *Myrmecocystus mexicanus*, an ant that lives in dry areas in North America must be mentioned. These ants store honeydew in living "honey pots," the specialized workers, or "repletes." These ants suspend themselves from the roof of the galleries of the ant nest and store honey in their stomachs, up to eight times their weight. This honeydew comes partly from aphids, but mostly from secretions from galls on oaks caused by certain species of gall wasps. This is a very specialized relationship between galls and ants. We also see that galls, while providing shelter and food for one species, can also supply surplus food for others.

In those galls produced by mites, mostly species of Eriophilidae, it is very difficult to tell where the gall starts and the acrodomatia ends. The responsibility of the mites, as we have seen in a previous chapter, is not always evident in the production of the various lodgings. Sometimes there are prefabricated domatia produced by the plant.

We will treat here essentially those galls caused by insects, particularly those induced by Coleoptera (weevils), Hymenoptera (Tenthredinidae, Cynipidae), Lepidoptera (Tortricidae), Thysanoptera, Homoptera (Aphidiidae and Coccidae), and Diptera (Cecidomyidae). The subject has been dealt with so often in the older literature, as well as in recent papers that we are able to summarize the subject. Sometimes we forget that the large Sagrinae (Chrysomelidae), with their bright colors and enormous femora, are gallicolous at the larval and pupal stage mostly inside the stems of big species of Leguminosae. They are a bit like the bruchids (Coleoptera) to which they are related. They do not, however, infest seeds. The archaic genera, sagrines are pantropical insects, dull brown in color, and exhibit a Gondwanian distribution (Australia, Madagascar, Argentina, Brazil). Old genera, as well as the recently evolved *Sagra* and related genera, have a unique biology among the Chrysomelidae. They form a simple type of gall which is a transitional structure, between mines and true galls. As mentioned above, the relationship between the sagrines and the bruchids (the latter strictly spermophagous beetles) cannot be overlooked. It is possible that the gallicolous insects were once leaf miners and later were "captured" and "maintained" by the plants to limit their damage. The hypothesis is attractive and could explain the behaviors of many larvae.

The gall is certainly a tumor, but a mild (benign) form. Generally the gall does not grow before the larva inside hatches and starts eating the surrounding tissues. The growth of the gall stops when the larva becomes a pupa.

Galls may be swellings, cupules, and spheres. The insect egg, and the larva which hatches later, can be either inside plant tissue or at its surface. Each combination of insect and plant produces a typically different kind of gall.

Certain galls are well known: the pineapple galls of spruce (*Picea* spp.), with swollen leaves overlapping each other, those produced by adelgid aphids *(Adelges laricis* and *Sacchiphantes abietis)*, the oak galls made by a cecidomyid fly, the round galls (apple galls) of oak made by a cynipid wasp, and the hairy galls of roses produced by another cynipid. Galls are called tumors because of the way they are produced by an increasing growth of tissue, a cell hypertrophy, a nucleus alteration, and a more rapid division of the cells.

Experiments have shown that gallmaking is due to the introduction of an auxinlike substance produced by the animal, generally the newly hatched larva. This larva secretes the substance through its skin (as in *Mikiola* sp.), or via its salivary glands, during the time it feeds on plant tissue. An exception is the species of *Pontania* (Hymenoptera, Tenthredinae) and gallicolous species of *Salix*. The galligenous substances are produced at the adult stage, during egg laying. Other more elaborate hypotheses about gall production are also being studied.

Recent research on the relationships between a cecidomid fly, *Wachtliella persicariae*, and *Polygonum amphibium*, a plant of humid surroundings, have shown clearly that the tissue growth making the gall is entirely due to the feeding activities of the insect (Dieleman, 1969). This comes about by a change in the normal development of the leaf, caused by the increase of auxins, growth substances in the gall tissue. This gall is necessary for the insect development.

An old experiment by Molliard, who inoculated healthy poppy pistils with extracts of squashed *Aulax papaveris* larvae, resulted in inducing the beginning of gall formation. Although begun, it did not continue due to the lack of continuity of the auxin secretion. The biochemistry of the galligenous process remains obscure.

Saliva of species of the aphid genus *Phylloxera,* a sad memory in Europe because of its damage to grapes, contains numerous substances which are supposed to accelerate or inhibit growth, including various amino acids. Normally that growth is produced only if the galligenous substances contact young cells or tissues. But in the case of galls produced on the leaves of roses, in organs already developed, the cells return to the embryonic state, a bit like carcinogenic cells. If we remove a larva from its gall, the gall development stops immediately, the plant does not react any longer to the inducive substance.

The number of galls in the world is enormous. Many in the tropics have never been enumerated. In Europe, 200 kinds of galls have been recorded on three species of oak, cataloged according to morphology, location, and inducing insects. Oak apples are well known and have a pleasant appearance, first yellow, and later turning red. These galls are always situated on the margin of the lower surface of the leaf.

The histological structure of galls is variable, sometimes rather complex. There is an epidermis, with or without stomata, and often also with hairs which do not exist on the original epidermis of the unaffected parts of the plant. This epidermis covers an abundant parenchyma which is sometimes a bit sclerified. The plant organ (a leaf for instance) is entirely modified.

The gall is also provided with abnormal phloem and xylem vessels which may be partly ligneous. Sometimes a cavity is found inside the gall close to or in communication with the outside, and inside this cavity lives the parasite. Near this may be found a layer of hypertrophied nutritive cells with fragile membranes and incised nuclei. This layer feeds the larva. The cells start to regenerate as the larva feeds. Then this layer is surrounded by a solid layer, somewhat ligneous and, hence, protective. Tannin inside the gall produced by *Cynips* spp. on oak is situated in the external parenchyma tissue. It has also a protective as well as a repellent role.

Roughly speaking, the oak gall appears to be most profitable for the cynipids which live inside. The ligneous layer and the tannin protect the larvae against predators and somewhat against parasitoids, particularly the ichneumonids, although rather poorly, since the parasitoids are numerous and persistent.

The growth of some galls is so well organized that they open automatically at the right time for their inhabitants to escape (as in the moth, *Cecidoses eremitus*). Only insects with chewing mouthparts can dig galleries in order to escape, but many other ingenious systems have evolved, as for example, the piercing pupa of a species of the fly genus *Giraudiella*, living in the common reed, *Phragmites communis*. Gall aphids escape the gall only after the leaf wall dries.

The chemical composition of the galls has been well studied and several of them are very rich in tannins and in anthocyanins which give them their colors. Organic acids and glucids are important within galls and a valuable comparison has been made between galls and fruits. Both formations seem to be vegetative organs similarly modified, one by the larval parasite, the other by the fertilized seed, a parasite not unlike an animal embryo in its mother's womb. In both cases, the plant's reactions are analogous to an animal's, it supplies the necessary food and chemicals necessary for growth. A similar inducive change can be caused by a mycoplasm parasite, which can modify completely the aspect of a flower transforming all the flower's components (petals, sepals, stamens, pistils, and ovary) into green leaves. It is this phenomenon of phyllody which proves that many living organisms can modify the directions coded in the genes of cell chromosomes.

Many galligenous Hymenoptera show a complex cycle similar to that of aphids, *i.e.*, they have an alternation of generations. For instance, *Andricus quercuscalicis* produces galls on oak acorns (*Quercus robur* [fig. 15]) and the following spring, parthenogenetic females escape from the gall. These females produce, in turn, very small galls on the stamens of *Quercus cerris*. Both males and females are produced in the galls and the female later lays eggs in the acorns of the first oak.

A more complicated case is that of the pineapple galls formed by the spruce aphid. Their cycle takes two years to complete and comprises two generations. Also, the synchronization of the opening of the gall with insect development is remarkable.

In summary, galls have a certain morphological and chemical analogy to fruit. They seem the best solution for the plant in order to limit the damage and the displacement by the parasites.

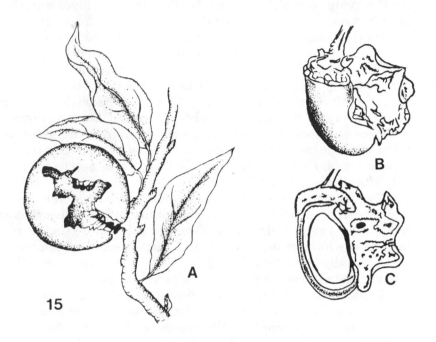

Fig. 15. Galls: a.Gall nut inhabited by the ant, *Colobiopsis truncatus* (after A.Raigner, 1952). b. Gall made by the wasp, *Andricus quercus calicis* on an acorn of the oak, *Quercus ruber*. c.The same in transverse section with the larval cavity shown (after Schremmer, 1976).

"That is well," Candide
answered, "but I need to take care of
my own garden ... "

- Candide, Voltaire.

CHAPTER 10

Fungus Gardens

Insects and fungi are often closely associated, but in certain rather well known cases, some insects actively grow specific fungi, feed on them, and control both their evolution and their growth. Generally, but not always, these phenomena are linked to the subsocial or social life of the insect.

In Greek mythology, the gods were eating nectar and ambrosia, which brought them immortality:

"At the table where one is never fully satisfied,
They were drinking nectar and eating ambrosia."

Thus wrote Victor Hugo. The botanists use the name nectar for the sweet juice of the plants, mostly produced by flowers, and ambrosia (from *Ambrosia* spp.) for the fungus covering the walls of the galleries of certain xylophagous Coleoptera (Scolytidae and Platypodidae). Anglo-Saxon writers have, therefore, named "Ambrosia Beetles," the various Coleoptera feeding on these fungi. Several ants, and certain termites, share this kind of food. Each of these insects grows its own "ambrosia," which, if not insuring immortality, at least provides them with a permanently available food and perhaps with the vitamins and chemicals necessary for the digestion of cellulose for those termites without symbionts.

Before talking about the real fungus growing specialists, *Atta* ants, the famous leafcutters, and about certain termites, let us review some of the amateurs and semiprofessionals among the fungus growers. The advantage of fungus farming is that a plant, devoid of chlorophyll, is easily cultivated in darkness and lives on organic matter which, in decay, transfers energy.

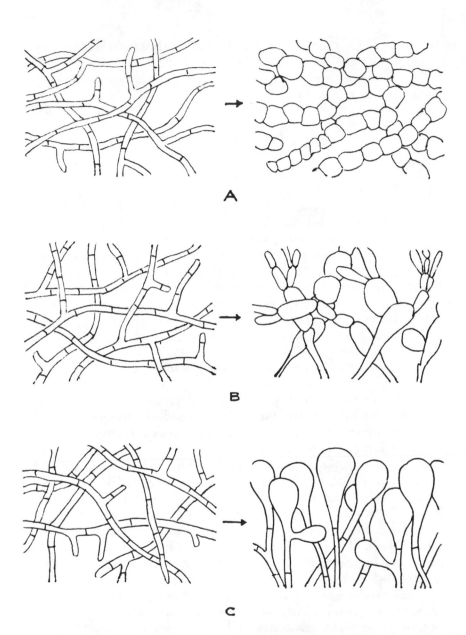

FIG. 16. Changes of a fungus after being cultivated by insects. On the left, fungi grown in Petri dishes. On the right, fungi improved by insect cultivation as shown by the enlargement of spherulation. a. Coleoptera (Scolytidae). b. Termites. C. Ants. (After Batra, 1967).

We are not dealing here with insect associations with internal symbionts: bacteria, yeasts, and protozoa. These aid insects in the digestion of their cellulose food or provide them with nitrates or vitamins. Our interest is only in those insect associations, nearly symbiotic as well, with fungi inside their nests (Batra, 1967-1979).

It is true that many insects control the growth of their fungi (Fig. 16), according to their own needs. Often, these fungi do not exist alone in nature. That control certainly results from the use of antibiotics produced by the insect itself.

Let us enumerate, one by one, the various categories of gardeners, beginning with the amateurs:

Galligenous insects

We just studied the phenomenon of gall making, the abnormal plant tissue produced by eggs and larvae of insects laid in a bud, leaf, or stem of a plant. But we did not mention that galls of itominid flies contain fungi which grow on food tissue and form a thick layer inside the wall of the gall. This fungus seems to be relatively specific for each dipterous larva, but it also grows independently.

Apparently it is the female fly which introduces spores of fungus when she lays eggs. It is also probable that the larva does not feed on the fungus itself but rather on the remains of the digestion of internal tissue of the gall by the cryptogam. Very few of these fungi have been properly identified. They belong to various genera such as *Sclerotium, Diplodia,* and others.

Coccidae

Species of several coccids (scale insects) of the genus *Aspidiotus* (Homoptera) live within the fungal tissue of species of the genus *Septobasidium* found on trees. The insects cause a modification in the structure of the fungus that it colonizes and this protects them from predators, and parasitoid infestations. The scale insect feeds on the sap of trees, but some individuals of the colony are literally sucked by the hausteria or threads of the fungus, providing it with nourishment. Hence, some scale insects perish, for the benefit of the colony, which is literally living inside a living shelter. The fungus does not exist without the insect: it is distributed by them and fed by them.

Hymenoptera

The larvae of species of the wasp genus *Sirex* (Hymenoptera), and related sawflies, are associated inside their wooden galleries with several fungi of the genera *Stereum and Daedalea.* The female wasp lays eggs in wet wood with her long ovipositor. The ovipositor contains at its base several cells (oidia) of the fungus stored inside minute cavities. The oidia attach themselves to the eggs and the mycelium grows from that germ and later covers the inside of the galleries. The fungus is not absolutely dependent on the insect, nor the insect on it, but it seems to help the larvae digest wood. Hence, we are not really speaking about a true symbiotic hymenopterous larva, but at least the future female also carries the fungus germs in very tiny waxy cavities situated between the first and second abdominal segments. It is only when the females hatch from the pupae that small pieces of fungus migrate into the base of the ovipositors and transform into oidia.

Coleoptera

The woodboring beetles (Scolytidae, Platypodidae, and Lymexylidae), most of which are called "ambrosia beetles," cannot survive without their specialized fungi (Ascomycetes and Fungi Imperfecti; the true identity of the latter is slowly being discovered). These beetles carry fungus spores in pockets (the mycangia [fig. 17]) situated at the base of the anterior legs of adults. The spores rapidly germinate inside tunnels carved into wood, mostly freshly cut wood, rich in sap. The beetles, larvae and adults, feed on fungi rather than on the wood itself. The smell of yeast attracts adults, as may be seen when they fly around in the evening near empty beer cans. The tunnels dug by these insects contain adults of both sexes, larvae, and eggs, and have a characteristic shape and color. Also, there is a close association between a beetle species and "its" fungus. It is the formation of yeast which is prevalent inside the tunnels and that is influenced and maintained by the insect.

Ants

Ants are highly evolved, social Hymenoptera. Some tropical ants in the New World, belonging to the subfamily Myrmicinae, especially species of the genera *Atta, Trachmyrmex,* and *Cyphomyrmex,* among others, use a sort of manure (compost), made chiefly of cut leaves, on which they plant a fungus. The fungus is eaten and provides the only food of these ants. The 400 known fungus growing ant species belong to only one tribe, Attinae, and these are mostly species of *Atta,* the best known genus.

Only in the 20th Century was it ascertained that the leaves carried by these ants, held over their backs (hence, the parasol ants [fig. 18]), were not used directly as food or as nest building

material. Instead, the leaves are used to fertilize underground
fungus gardens.

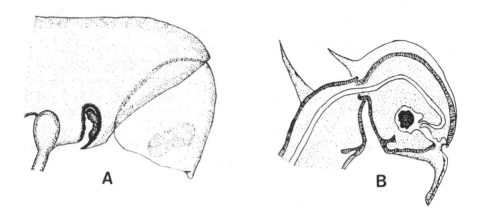

FIG. 17. Fungus cavities in beetles: a.A scolytid beetle in longitudinal section
carrying fungus spores in pockets situated at the base of the anterior legs. b.
Section through the head of a fungus eating ant showing the fungus pocket.
(After Batra and Batra, 1967).

When the very large *Atta* female is fertilized, she drops her
wings and looks for a nest in the ground. She digs a cavity in the
soil and remains in it. She has carried into the soil material collected on her body and stored in an infabuccal pocket. This material, planted in the soil cavity, starts the fungus garden which is
going to feed the entire colony. In this material are hyphae, or
threads, from the previous meal of the future queen and these
hyphae will act as seeds to start the garden. For many more days
the queen will fertilize these developing fungi with her own excreta, utilizing even some of her own crushed eggs both as manure
and for food. She constantly cultivates the garden with her
antennae and legs. Some eggs which are laid become workers, fed
when in the larval stage, with the eggs of the queen, then, when
adults, they feed on fungus which they will grow themselves after
collecting leaves. Excreta of caterpillars, plant debris, and anthers from flowers are also collected by some ants to improve the
quality of the compost or replace the leaves. Generally this is
done by small colonies.

The leaves, cut and carried by the workers to the nest, are
ground, mixed with saliva, and covered with excreta. On this

compost, the ants place some mycelia to seed the garden in their nest. After a certain time, the total fungal mass will resemble a sponge, rather similar to a termite garden.

It is to be noted that the transportation of fungal spores is not the only material carried by the queen ants. For instance, a tropical ant, *Acropyga paramaribensis,* rears up to five different species of Coccidae on roots in the nest. Each virgin queen that leaves to start a new colony, carries a young female coccid in her mouth during the nuptial flight. This adaptation is not as specialized as it is for the growing of fungi but it produces the same end: to perpetuate the food of the species. On the other hand, the virgin queens of species of *Carebara* carry, during their nuptial flight, very minute workers attached to their legs.

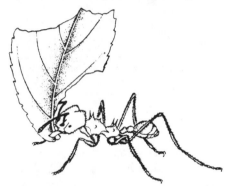

FIG. 18. a.Leaf cutting ant carrying a piece of leaf to the nest (after Linsenmaier, 1972, and Dumpert, 1978).

Atta ants are not the only leaf cutters. Species of *Megachile,* solitary bees, are well known for their ability to cut leaves to fill the inside of their nest which is a cylindrical burrow. These bees have dentate mandibles, which, in the tropics, they use to cut leaves of species of *Eucalyptus* wherever those trees have been introduced. However *Megachile* bees do not cultivate fungi. The cut leaves serve only as a protective device within the underground nests.

Under the constant care of workers, the *Atta* garden takes shape. The fungi produce special terminal bulbs on the hyphae which have been called bromatia, ambrosia, or kohlrabies (fig. 19). Only the ant gardens produce this nonreproductive element, the ant's food. Bromatia do not appear spontaneously in culture media. Ant larvae are placed on these fungus gardens and it is here that all of the members of the colony feed.

The nests are so large that sometimes a bulldozer is necessary to break them up, before construction can begin on the land.
Underground, *Atta* nests can reach a considerable size, 4.5 to 6 m
as well. These chambers may measure one meter long, 30 cm
high, and 30 cm wide, but that is the maximum. Some species
build smaller nests. The craters of the nest and the naked soil
around it resemble the outside of the nests of other ants, particularly the harvester ants. A nest can cover one hundred square
meters and possess hundreds of entryways. *Atta* workers move in
columns to cut leaves and their enormous numbers can be very
destructive, sometimes completely defoliating a tree in a very
short time. Workers, from a well established *Atta* colony, can
defoliate a tree in one night and then return when new leaves
sprout. As protection from the voracious *Atta* ants, some trees,
especially species of *Acacias*, have either developed toxic substances in their tissue, or myrmecophilous associations with other
species.

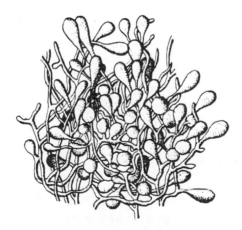

FIG. 19. Ambrosia or bromatia of a fungus cultivated by a fungus eating ant
(after Wheeler, 1923, and Dumpert, 1967).

A large nest may contain a hundred gardens, and some
among them are suspended from roots in the ceiling of the
chamber, but such cases are rather exceptional. The temperature
and humidity of the nest is very important and ants keep it
constant by opening and closing nest entries. Humid dew covered
leaves are not collected, therefore, much of the harvesting is done

at night when they are dry.

As is frequently the case among fungus eaters, each species of ant cultivates its own species of fungus and it is difficult to make them change this artificially. These fungi are specialized for the ant species and the ants probably have attained the most evolved state of gardening, better, in certain aspects, than those of the termites, which the ants have imitated only by convergence. When ants do not eat bromatia, they lick them for their secretions. These bromatia look like the white spherules of termite gardens where larvae exclusively feed on them.

As long as ants cultivate their garden, fungus does not sporulate or produce fruiting bodies. However, the species of four genera of these cryptogams have been cultivated or found in open air after the withdrawal of ants from their nest. They do sporulate when unattended by ants. How are the bromatia produced? Probably by constant pruning of mycelia and by antibiotic effect of ant saliva and excreta. These antibiotics not only influence growth of symbiotic fungi, but also prevent germination of other species of cryptogams.

The *Atta* colony is polymorphic. Minute workers, the "minimes" almost exclusively cultivate fungus and feed the larvae. A second form, the "medium" workers collect leaves. A third form, large ants, are the last to develop and have mainly a defensive function similar to the soldiers of other species.

The colonies of fungus ants vary considerably in their organization. Some species make only one garden (species of *Acromyrmex);* others have only one type of worker, and still others, have queens who also collect leaves. Some fungus gardens are covered by an envelope, a part of the fungus itself.

Termites

Fungus growing termites, in contrast to the ants, are found only in the Old World tropics and all belong to the single family Macrotermitidae. Species of *Macrotermes* and *Odontotermes* make their nests inside high, sturdy mounds. Other species, such as those of *Microtermes,* are subterranean and the nest is only slightly visible. The biology of these termites is well known. We will only treat the general aspects of the subject here.

Organic matter, mostly wood collected, chewed, and digested by the workers, makes up the fungus gardens of the termites. Sometimes there is a single garden, which can weigh up to 30 kg, but generally there are numerous gardens dispersed throughout the nest. The inner cavity of the nest is lined with a mixture of saliva and dust. It is sometimes ventilated by a system of vertical conduits reaching the outside. The gardens have a spongy appearance and are grayish or brownish in color. They are kept humid. Some gardens are firm, others soft and fragile. On the surface of the gardens are seen small, white, shiny, pearlshaped spherules. These are composed of rounded cell masses, the "mycotetes," bundles of threads emanating from the surface of the mycelium strands of the garden. These whitish spherules are collected by workers and sometimes eaten or offered to certain nymphs. It does not seem that the king, queen, and young nymphs eat fungus spherules, but the alate and reproductive forms take this fungal material with them in order to seed new fungus gardens.

Contrary to most termites of the temperate and tropical areas which harbor intestine protozoa for digesting cellulose, no protozoa seem to be living inside the intestine of fungus growing termites. From this it has been deduced that cultivated fungi predigest cellulose and provide vitamins. These termites cannot live on wood and continuously graze on the garden material. They use their own fecal matter to fertilize the fungus which is then eaten, and thus recycled.

Several genera of fungi grow inside termite nests, but species of *Termitomyces* are the most common and these cannot be found elsewhere.

Fruiting bodies do not generally appear when the termites are inhabiting the nest. However, under certain favorable conditions, *Termitomyces* species reproduce sexually. The spherules grow, reproduce, produce carpophores with strong caps. When they get out of the nest and mature, they disseminate spores. *Termitomyces* species have been studied in detail by P. P. Grassé and Roger Heim who have determined that these species are closely related to *Agaricus* species (Basidiomycetes).

In Africa and India termites sometimes remove the center layers of their gardens and spread them outside the nest during the rainy season. Soon "mushrooms" (the spore bearing caps) appear. In this manner the spores of the various nests are mixed together. The termites then collect the mycelia produced from this mixture of spores. It seems that in doing this the termites bring about a cross fertilization of their fungi.

We may conclude that there is a close association between ants and termites and their respective fungi. Fungi produce food for their hosts, from their substratum, and the insects eat either fungus, the substratum predigested by the fungus, or both. Although fungus is prevented from producing reproductive structures, it receives from its insect host a safe ecological niche and often a means of dispersal.

The situation is reversed in certain coccids. In this instance, the insect is sheltered by species of the fungus genus *Septobasidium*. The scale insects feed on their plant substratum and provide food to the fungus from their own bodies.

In all cases examined, there is a more or less perfect symbiotic relationship. First, a dissemination system for the fungus is provided. This may be simple, or sometimes elaborate. Second, it is certain that, due to antibiotics secreted by these insects, only fungi useful to the host are allowed to grow; all others are totally inhibited. Such substances are also partially responsible for the fungus' metamorphosis which, without its gardeners, does not produce bromatia, spherules, or ambrosia, but rather retains its archaic form of reproductive body. Until new findings show otherwise, the saliva and excreta of ants and termites, are presumed partially responsible for these changes. In the case of beetles, however, there are no spherules, no bromatia, but the cryptogams are transformed into yeast. It also seems that temperature conditions, quantity of carbon dioxide in the air, acidity, and so on produce similar modifications in the fungus and that the phenomenon is not produced only by the insects. Laboratory trials are not absolutely conclusive, however, and much remains to be found and to be proven in this fascinating field.

"It will rain bread from the sky.
When the Israelites saw it,
They said to one another,
This is the manna which the
Lord has given you to eat.
It was like coriander sod,
White, and it tasted like
Wafers in honey.
The Israelites ate manna for
forty years. . . "

-Exodus XVI, from verse 16.

CHAPTER 11

Entomological Manna

Manna, the sweet secretions from plants, or honeydew excretions from insects, is well known chiefly because it was the "bread of heaven" which allowed the Israelites to survive for forty years in the Sinai Desert. Honeydew production by aphids, cochineal scales, and psyllids, or sometimes by cigal leafhoppers (all Homoptera), is a good example of adaptive symbiosis (as opposed to parasitism) between plants and insects. Other interpretations of the Israelite manna include the lichen hypothesis. Manna could have been that of the lichen, *Lecanora esculenta,* which produces granules of starch which may be spread over considerable distances by the wind.   Bedouins in Libya make flour and bread from that type of vegetable manna, which is rather indigestible because the lichen contains oxalic acid. Let us say that this hypothesis, although a rather tempting one, seems doubtful, since after one and a half centuries of searching no *Lecanora esculenta* has been found in the Sinai Desert. However, a combination of lichen and insects could reasonably explain, at least in part, the Biblical episode cited at the beginning of this chapter.

Two scale insects, one in the Sinai mountains, *Trabutina mannipara,* and one in the plains, *Najacoccus serpentinus,* pierce and feed on the branches of tamarisk trees, *Tamarix mannifera* and *T. orientalis,* and then excrete a sweet liquid which falls to the ground and hardens into small particles. These sometimes take the shape of the whitish flakes described in the Bible.

Bedouins collect this honeydew, considering it a great delicacy. The salt which sometimes exudes from the leaves of tamarisks, essentially halophytic plants, depreciates the manna quality.

The honeydew produced by aphids is a mixture of glucose, sucrose, fructose, and meleitose. Some large aphids can drain up to 1.4 mg of sugar per plant, per day. If you consider that the average duration of an aphid's life is about 30 days, that means a single aphid can draw from 30 to 120 mg of sugar. One million aphids on an average shrub can produce 1 kg/m of sugar in a summer, which is considerable. Additionally, the quantity of sugar produced by tamarisk scales is certainly much greater than that produced by the aphids.

Honeydew, instead of pollen, is known to be collected by bees, but they also collect from various kinds of flour found in African markets and these extrafloral collections seem, in fact, to be very local. Certain bees even collect completely indigestible brick dust, a mistake made because of instinct and their inability to distinguish this substance from pollen. The harvesting of honeydew, however, is probably an adaptation to the lack of flowers at times in desert areas. Conversely, ants take honeydew during the diurnal peak of their activity. This is one interpretation of the maggots mentioned in the Bible as destroying the manna stored in excess, despite divine orders not to do so: "And Moses said to them, 'Let no one leave any of it over until morning.' But they paid no attention to him; some of them left it until morning, and it became infested with maggots and stank." (Exodus XVI: 19.)

Many other scale insects secrete compounds useful to humans, including wax and shellac taken from various plants. Manna can be collected by the kilogram during rainy years in certain areas of central Sinai. Tamarisk manna is mostly rich in glucose, fructose, and saccharose. The droplet excreta of the nymphs or females of the cochineal scale is transformed into sticky granules by the dry air of the desert. According to Bodenheimer, the continual process of insects pumping sap from the tree is due to the necessity of absorbing sufficient nitrogenous compounds from these desert plants, which characteristically contain low levels. Tamarisk sap is known to be deficient in nitrogen. Owen thinks that this sugar excretion by the homopterans is generally beneficial to the plants themselves as it drains excess sugar and increases the nitrogen fixation in the ground when bacteria act on the glucose deposits. This hypothesis is far from proven, and it is in opposition to the traditional ideas of agronomists who consider aphids to be essentially pests of plants (for which they use insecticides). However, this beneficial view of aphids may well be correct.

Other manna, similar to or different from that of tamarisks' insect-plant associations, exists in other deserts, namely in Iran, Syria, and in the Middle East in general. In Iran, manna is excreted or produced by the nymphs of a psyllid, scale insects, leafhoppers, and aphids, but this is manna of a very different nature. Even cocoons of species of weevils of the genus *Larinus* found in the Kurdistan Mountains, when boiled, produce a trehalose manna. This sugar was named by Berthelot, in 1858, from the Kurdish name for the cocoon, Trehala. An oak aphid in the vicinity of the Urmiah salty lake (Rezaieh), in Kurdish Iran, produces a honeydew made of trehalose which, when mixed with flour, makes delicious cakes. The Urmish area, which we visited in 1956, is populated with Kurds and Christians of the Chaldrean rite (Ussuri or old Assyrians). Both communities have similar traditions and use natural manna found in their province. Both Persian manna and Kurdish manna, also called "manessimma" or heaven manna, are very well known throughout the region.

Finally, the association of these homopterans with plants is more than parasitism; it can be considered to be partially symbiotic. The production of honeydew is probably used by the plant to increase available nitrogen, and the insect, in turn, is able to extract for its own use the necessary chemicals for its metabolism.

"A fallen flower
From its branch, I see it
coming back.
It's a butterfly."

-A. Moritake, XVth Century.

Chapter 12

Pollinating Insects

The flower is the product, indeed, the symbol, of the last group of plants to have evolved. It is certain that the flower is a structure which combines a protective envelope around the reproductive organs, the sexual cells, the pollen and ovules. Flower parts are only modified leaves. This can easily be seen when a plant is parasitized by mycoplasma, a type of organism combining features of both bacteria and viruses, but with enough distinct characteristics to be assigned to a separate plant division of the bacteria, the Mycoplasmas. The flower, whether compound like those of sunflowers, or simple like those of cotton, transform partially or entirely into a leafy organ when infected by mycoplasmas. Petals, sepals, stamens, and pistils become green leaves. This phenomenon is called phyllody and can sometimes produce rather esthetic effects, for example, those of cotton, but of course, the parts are totally sterile.

The hermaphroditic flower of many Angiosperms is composed of the perianth, or floral envelope, made up of the calyx (sepals) and the corola (petals), the androecium (stamens with their anthers) and the gynoecium (pistil and ovary). At the base of the pistil is the ovary or the ovaries from which emerge the styles capped by the stigma. Pollination, the prelude to the fecundation or union of the male and female gametes, is accomplished by transportation of pollen grains from the anther of the stamen to the stigma (in angiosperm flowers). In the gymnosperms, pollen from the anther is taken directly to the ovule in the cone. The fusion of the gametes in the ovary will produce an egg which will develop into an embryo and, along with others parts, will form a seed. Insects and wind are the main pollinators of plants, helped by other mechanical means such as water in certain primitive flowers.

Not only do insects pollinate plants but some special adaptations allow birds, such as hummingbirds (Trochilidae) in America; sunbirds (Nectarinidae) in Africa, Southeastern Asia, and eastern Australia, sugarbirds (Meliphagidae and Trichoglossidae) in the Australian region, the Hawaiian Honeycreepers (Drepanidae) in Hawaii, and bats in tropical America, Africa, and Asia, to effect this process. At least one flower is pollinated by rats in Hawaii (*Freycinetia* sp.). Species of *Banksia* in Australia are pollinated by small marsupials. Their long tongues and fine fur is covered with pollen which is then transported from plant to plant. Some snails pollinate certain flowers (such as species of *Aspidistra* and *Rhodea*). Orchids are pollinated, according to genus and geographical areas, by Lepidoptera, Hymenoptera, birds, bats, frogs, slugs, and for two endogenous Australian species, by earthworms.

Pollination methods are named as follows: anemogamy, by wind; hydrogamy, by water; entomogamy, by insects; ornithogamy, by birds; therogamy, by mammals; and cheirogamy, by bats. Therogamy is usually restricted to rodents and marsupials. The Proteaceae (which includes the macadamia nut) in South Africa, are remarkably morphologically and biologically adapted to "their" rodents by having an abundance of nectar, a special smell and color compatible with the nocturnal life of the animal, strength to hold their weight, and a large flower and petiole near the ground. This coadaptation is almost as perfect as those between insects and flowers described on the following pages. There is also a temporal coincidence between the time of flowering and the nutritive needs of the animal. The same adaptation also occurs among the Australian Proteaceae and their marsupials, and must exist elsewhere, namely with primates in the tropics (for example, *Galago* sp. with species of *Adansonia*, *Loris*, and *Tarsius*). If these facts are not well known, it is due to the difficulties in observing these nocturnal, shy, and rare mammals. It does not seem that pollination by the vertebrates developed later than the insect-plant syndrome. On one hand, there exist fossil traces of presumed relationships between insects and plants, and, on the other hand, the adaptation of the South African Proteaceae to their rodents is too specialized to be fortuitous.

There are not any pollinating birds in Europe or temperate Asia, perhaps the reason for the rarity of red flowers in these countries. Most insects, except pierid butterflies, are blind to red.

Wild red poppies actually reflect ultra violet, at least as far as bees are concerned, because they do not see the same colors as we do.

Hummingbirds are often migratory. In summer they fly as far north as Alaska, where they find flowers adapted for pollination by them. Their flowering time is adjusted to the birds' migrations. One African nectarinid bird flies north to Palestine where red flowers of the epiphytic mistletoe, *Loranthus acaciae,* are perfectly adapted for pollination by these birds. Some species of *Fuchsia,* growing in the extreme southern tip of South America, are pollinated by birds. Cultivated species of *Fuchsia* which have become nearly wild on the island of La Reunion, near Madagascar, have no birds adapted for their pollination. They seem to reproduce only vegetatively. At high altitudes, as in the Andes, a similar adaptation occurs with the bromeliad, *Puya raimondii,* in the East African mountains with blue flowering giant *Lobelia* blossoms, and in the Papua New Guinean mountains with species of *Rhododendron.* All of these bird-plant associations are specialized and indicate parallel evolution.

During a stay in eastern Papua New Guinea, I was impressed during the climb of the highest peak, Mount Wilhelm, by the diversity of size, shape, and color of the rhodendrons. There, a special adaptation seems to exist between the pollinators and the flowers: at higher altitudes, red and odorless flowers are fertilized during the day by birds and at lower altitudes, white and smelly flowers are fertilized at night by Lepidoptera, mainly hawk moths.

It must be noted that sometimes, as in tropical America, wherever Lepidoptera and hummingbirds coexist, both become competitors chasing each other. Hawk moths are chased by the birds. At high altitudes the moth, *Castnia eudesmia,* chases the birds away from the flowers of *Puya alpestris.* The bird is often smaller than the moth in this tropical forest. Not only do the nectarinians fertilize the flowers but there is a well known case of a species of *Boerlagiodendron* (Araliaceae) attracting pigeons. This plant has fruitlike bodies which, according to Van der Pijl, are actually sterile flowers. It is difficult to know whether the nectariphagous birds are primary or secondary pollinators because they search for insects, water, food, and nectar. However, the adaptations seem very old and probably began at the end of the Jurassic with the appearance of both the birds and the first flowering plants.

The distribution of bats, which are color blind, and the plants they pollinate, all of which smell like ripe fruit, coincides perfectly. The adaptations seem to be very specialized and exclusive.

Hummingbirds and hawk moths do not land on flowers when they are gathering honey, but hover like helicopters with their elongated beaks or proboscises extended. Nectariphagous bats show adaptations: elongated snouts, and elongated tongues with corneous expansions at the end. Insects also have similar adaptations, resulting in a great diversity of sucking organs for gathering nectar. Many plants are dependent exclusively on certain insects, as well as birds and bats for fertilization.

Many pollinators fly high in the canopy of tropical forests. These include insects, bats, and birds. Among the insects are some Diptera and Orthoptera, all polyphagous, but this specialized fauna is rather poorly known. Scattered throughout this canopy are trees with large flowers, easily visible from a plane, which break up the green monotony of the canopy. Some forests exhibit more color than others, according to the season, the geographical area, the local flora, and the altitude. At the top of the canopy, where the temperature reaches 40° C, flowers are numerous and pollinators are abundant. At the same time it is evident that pollinators at the lowest level of the tropical forest are rare because the shade cuts down on the number of available species. At the soil level, many flowers are cleistogamous or endogenous (*e.g., Commelina* spp.).

As previously mentioned, honey gathered from *Rhododendrons* (with about 1200 species) is toxic to humans, chiefly the species occurring in Asia Minor, but also elsewhere, including recent reports from Nepal. The honey is not toxic to bees, but numerous plants produce a pollen which kills bees, or at least makes them sick. The pollen of buttercups (Rununculaceae) and of some shade trees is poisonous. Normally, bees avoid these plants, but often when food is scarce, or during a drought, bees will gather toxic pollen.

Much has been written on pollination and pollinators, the specialized associations between plant and insect, and the many adaptations which have evolved. Pseudocopulation of certain Hymenoptera with orchids is a relatively recent discovery which has been confirmed to exist almost everywhere, even in Australia.

The biology of bees is now well known and described in almost all biology textbooks. Their relationships with the flowers they pollinate are now known to be functions stimulated by smell, color, shape, and movement of the host plant. A motionless flower certainly attracts fewer insects than a flower shaken by the wind. Too much has been written on bees to go into great detail here.

Also we have discussed previously (Chapter 7) myrmecophilic plants. We should add here the fact that the Italian form of *Apis mellifica*, introduced into Mexico long ago, quickly became a rival of the ants that protect *Acacia* trees. The nectar produced by the trees attracts bees, but the bee is always the loser in case of conflict with the ants. The latter remain the most ferocious, since they were the first to become established on the tree and are defending their own territory (Janzen, 1975).

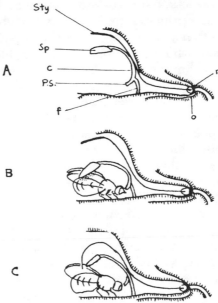

FIG. 20. Pollination of the flowers of *Salvia glutinosa* by bees. a. **f**, filament; **c**, connector; **PS**, palette; **sp**, anther; **sty**, style; **n**, nectaries; **o**, ovary. b. Flower at the male (pollen) stage; note pollen on bee. c. Flower at the female (ripe ovary) stage. (After Guillermond and Mangenot).

Cross-pollination is usually a necessity for most plants. Figure 20 shows how a bee pollinates the flowers of *Salvia glutinosa* (Jupiter's Distaff). Note how the anther bends to drop pollen on the visiting bee. Later, another bee, laden with pollen, pollinates the stype (c). Almost all Leguminous fodder plants are autosterile and this is also true of many fruit trees. One third of our food depends directly or indirectly on pollinating insects. Alfalfa, *Medicago sativa,* originated in the Middle East, but it is not of much interest to honey bees in our latitudes. In Europe, solitary bees, including the American species, *Megachile pacifica,* seem more efficient as pollinators than European species. Without pollinating insects, only 1% of the alfalfa flowers set seed. Also note that pollinators in Western Europe which nest in hedges fall victim to the destruction of hedges as well as the blind use of certain insecticides. Some races of honey bees are more efficient than others as pollinators. This can vary from 20% to 80% in the Caucasian variety which has a longer tongue than some of the others. This is probably why, in the middle of summer, and also in irrigated areas of Iran and Afghanistan, where alfalfa is native, it is pollinated by local wild bees and bumble bees.

A well known example of the need for entogamy is the vanilla plant, a climbing Mexican orchid, which has been introduced into Madagascar and La Reunion. Only species of American bees of the genus *Melipona* are able to perform the pollination, although thanks to a discovery by a Reunion slave named Albius, it is now possible to artificially pollinate vanilla flowers, which is absolutely necessary for the formation of the pod. He did this through the process of autogamy or self-fertilization.

Later in this chapter we will discuss many similar examples, including that of *Yucca* and *Ficus,* which, though quite common, represent a specialized and remarkable instance of adaptation. Fertilization of various species of *Ficus* is extremely diversified. Very few date trees are parthenogenetic; cross-fertilization is necessary. Fertilization is normally effected by windblown pollen from male trees, which reaches a certain proportion of female trees. In order to increase the chances of successful fertilization, someone must climb up the female trunk and suspend a male inflorescence near the female one. This practice is very old, at least since Babylonian times. We have observed this ancient custom in February (a dangerous operation because of the spines on the trunk) in the Sudan, along the Nile. Modern growers do this with a power sprayer using a mixture of dust and pollen. Thus, as with the vanilla plant previously mentioned, human innovation replaces insect and wind.

The discovery of sex in plants dates from the end of the 18th century. Despite the fact that numerous plants are sexually dimorphic, as date trees, with male and female plants, this is not very obvious. What visibly distinguishes a male papaya tree from a female, especially when male plants sometimes bear abortive fruits? Camerarius is honored as the first to distinguish between monoecious plants, with either hermaphrodite flowers or separate, unisexual flowers of both sexes, and dioecious plants with unisexual flowers on separate plants. He also discovered that stamens carry pollen, the male gametophyte, and that the pistil is the female organ. Through their combined action they produce fertile seeds.

Many books and journal articles have been written on cross pollination of flowers, especially by entomogamy. This chapter is an attempt to condense all of this into a few pages. Recent research on this in the tropics has enriched the older data acquired in temperate regions.

A fundamental law of plant fertilization is the inferiority of self-pollination and the advantage, or often, absolute necessity for cross-pollination, *i.e.*, the fusion of gametes produced by two different individuals. Sprengel, in 1773, and Darwin a century later, both insist that this is a fundamental biological law. If so, why are all plants not cross-fertilized? To answer this, let's see how these processes work.

The structure of angiosperm flowers reveals a neat tendency in favor of cross-pollination which makes self-pollination improbable, if not impossible. If it does happen, however, the resulting union of gametes is usually sterile. The means used by plants for successful reproduction are various; some of them are reviewed here.

Let us remember that flowering plants are most frequently: 1) hermaphroditic (flowers with stamens and fertile carpels); 2) others are diclinous monoecious (stamen and carpels on different flowers, *i.e.*, male and female flowers on the same plant); or 3) diclinous dioecious, with male and female flowers on different plants. Some well known examples of the third case are: pawpaw, date palm, willow, poplar, helm, and hop.

The pollen, which contains the male gametes produced by the flower's stamens, equivalent to the microspora of the cryptogams, can germinate only if dropped on the receptive surface of the stigma (or through the action of enzymes in the stomach of certain insects). This receptive surface is characterized by a papillose and sticky epidermis. The ovule located beneath the stigma in the ovary is the equivalent of the macrospore of lower plants.

Pollen may be classified according to the way it is distributed: anemophilous, by wind; hydrogamous, by water; or zoidogamous, by animals. Insects visit flowers. These are very well suited as pollinators because pollen is a high protein food eaten by both adults and larvae of many Hymenoptera, some Coleoptera and some Lepidoptera. Many insects also sip nectar (especially Lepidoptera and Diptera), a watery solution generally produced by insect-pollinated flowers, high in saccharose, sucrose, and fructose. Nectar exudes from nectaries, either from the receptacle surrounding the ovary, from the base of petals, from stamens, or sometimes from other special organs. It is always secreted by the flower. Of course, there also exists a quantity of extrafloral nectaries which are exploited by insects, as we have seen in the examples of myrmecophilous and carnivorous plants discussed previously. These nectaries only have an attraction value for the plant and do not play any role in the pollination process itself. Both the act of feeding on nectar, and the actual gathering of pollen, results in the transportation of pollen on the insect body. This is either as a disseminated powder, or by a special plant structure, the pollinium (a pollen sack which sticks to the insect). By visiting one blossom after another they carry pollen from flower to flower and effect the cross-fertilization of separate flowers.

All pollinophagous or nectariphagous arthropods are not, however, effective plant pollinators. There are, for instance, Hymenoptera and Coleoptera which pierce a hole in the corolla and "steal" their reward. Certain insects with a long proboscises suck up the nectar without touching the plant. Many other cases, including species of phoretic mites of the genus *Rhinoseius,* which are carried by hummingbirds, are nectariphagous. Their pollination potential is problematic. Other cases include many Thysanoptera (thrips) and some beetles. However, these are generally exceptions.

The classical example of insects that fertilize flowers, not because of nectar or pollen, or because of any other attractant such as perfume or nutritive tissue, but to lay their eggs in the ovaries of the plant, is the process of fertilization for both *Ficus carica* and species of *Yucca* flowers. Fig pollination is effected by a special galligenous hymenopteran species for each fig species. The best known is a species of the genus *Blastophaga,* a chalcidid wasp. The mature female bores into the fig inflorescence and lays her eggs in the ovaries of sterile flowers. Males produced from these eggs escape to fertilize female wasps. As they leave the male fig flowers they are dusted by pollen which is transported to other fig trees. Theoretically, the insect stimulates only the development of the endosperm in the ovules. It is this endosperm on which the larvae feed. In reality, the cycle is much more complex, since there are three types of flowers: male, female, and sterile.

A species of *Yucca*, a Mexican Agavaceae, is also remarkable in that it is one of the rare cases of an absolute mutual dependence between a flower and an insect. A small moth, *Pronuba yuccasella* (Tineidae), is involved. *Yucca* flowers give off perfume, mostly by night, and this attracts the moths. The female moth has a long ovipositor and is provided with prehensile, spiny, maxillary palpi specially modified in this genus. At night, with the aid of the palpi, the female collects a certain quantity of pollen, and shapes it into a ball about three times as big as its head. Then she flies to another flower, deposits some of her eggs in the ovary with her ovipositor. Next she climbs to the top of the flower and places the pollen ball on the stigma. The ovules are then fertilized, thanks to the remarkable instinct of the moth. In addition, pollen is so abundant that there is enough for larval food as well as for the reproduction of the plant. Without this method of pollination, the ovary would wither and the plant could not reproduce itself, which would mean the moth larvae would not receive their food. However, outside of America, namely in India, the autogamous (?) production of seeds in *Yucca alsifolia* has been described, but pollination has never been observed.

In temperate regions, species of *Mahonia* and *Berberis* (Berberidaceae) have stamen elaters (spring mechanisms) which provide them with mobility. These dust pollinating insects with pollen at the slightest contact. These movements, called "seismonastic," function similar to flower and leaf movement, carnivorous or not, seen in the sensitive plant *(Mimosa* spp.), certain orchids, and many tropical plants.

Self-fertilization is sometimes successfully realized in certain plants and this is necessary in certain plants with flowers which never open (cleistogamous). As yet unknown are the pollination methods of the underground, or geocarpic, flowers of certain fig trees, the saprophytous Australian orchids of the genera *Cryptanthemis* and *Rhizanthella*, . If cross-pollination, it must be carried out by earthworms or insect larvae. In these orchids, the flowers and the seeds are found 2 cm beneath the surface.

Cross-fertilization is aided by various anatomical structures and special physiological mechanisms. For dioecious flowers, cross-fertilization is obviously obligatory. In most hermaphroditic and monoecious flowers, cross-fertilization is also required as the pollen grains and the pistil of the same flower cannot unite because they are separated by time (dichogamy), space (hecogamy), or are incompatible (heterostyly and incompatibility). These examples are well known and explicated in the literature to which the reader is referred. Let us say that autogamy and self-fertilization are rare, and excepting the classical examples of certain plants, (some cereals, Papilionaceae, and those with cleistogamous flowers), it exists only where pollinators are rare (subdeserts, deserts, and cold mountains) or absent. Autogamy is often caused by the visit of the insects themselves.

When stamens of a hermaphroditic flower mature first, it is said to be protandrous. Conversely, if the pistil matures first, it is said to be protogynous. An example of the pistils and anthers spatially separated may be found in some orchids and milkweeds. A classic example, heterostyly, or production of flowers with a short style (brevistylous) and long stamens, or with a long style (longistylous) and short stamens, is to be found among the primroses. The specialized nature of pollen grains practically excludes, among those plants, any self-fertilization.

Hercogamy (the condition in which self-fertilization is impossible) or dichogamy (the condition in which the maturing of the sexual elements takes place at different times), as mentioned by Jaeger (in Grassé, 1976) favor cross-fertilization but the only absolutely sure system for these are dioecy (plants with separate sexes) or autoincompatibility.

It is evident that many insects are capable of pollinating plants ranging from giant tropical butterflies to minute Thysanoptera or Thysanura. However, Lepidoptera, Diptera, and Hymenoptera, are the most efficient and the most common pollinators.

Magnolias, usually pollinated by beetles, have strong female organs. Some beetles may be satisfied only with pollen, for example, scarabs (cetonids), but others often eat flower petals (some meloids and many Galerucinae). Rarely, other flowers are pollinated by beetles, such as the *Welwitschia* sp. from the Southwest African desert, by Tenebrionidae. The famous species of *Amorphophallus*, a giant flower close to *Arum*, is full of small beetles which feed on the rich oily and starchy tissues of these plants.

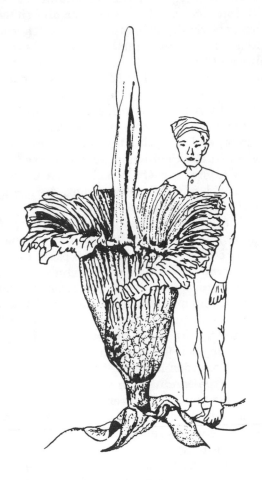

FIG. 21. Inflorecence of a giant Sumatran flower, *Amorphophallus titans* (Araceae). The red color and fetid odor of these flowers attract blow flies and induces them to lay their eggs on the spadix. Several species of beetles are attracted also. (After Meeuse, *in* Jaeger, 1976).

The archaic cycads have male inflorescences with an unpleasant odor; female flowers are fertilized by numerous beetles among which are Aulascoscelinae in tropical America, and species of *Phlaeophagus* in Austral-Africa. Several entomogenous gymnosperms have inflorescences with colored bracts (some species of *Ephedra)* which are visited by tenebrionids. Others (Cycads) produce food bodies.

Coleoptera are far from being primitive insects, but except for a few genera with a proboscis (some Meloidae), they are poorly specialized as honey gathering insects. There seems to be some correlation between Coleoptera and primitive flowers, but as we will see later, this is uncertain; we must remain careful and refrain from making generalizations.

Hymenoptera are the most efficient of all pollinators. The size and color of flowers are attractive to them; however, this is not equal for bees and humans. Their perceptions of pleasant or unpleasant odors may not agree. Flower bracts are colored in some plants, as for example, in species of *Poinsettia, Bougainvillea, Euphorbia,* and many others. In species of *Mussaenda,* one sepal is large, leafy, and brightly colored, and helps to make the flower conspicuous. Certain flowers have petals with spots or lines which are nectar guides. These guides are not always visible to the human eye, but sometimes appear in the ultra violet light range which is visible to bees. Humans see many contrasting colors, but bees apparently are unable to distinguish more than four colors: yellow, green-blue, blue, and ultra violet. This is a small number compared to the approximately fifty colors visible to humans.

Flower shape can also attract some insects. Some studies indicate that while bees show a preference for flowers with radial symmetry, bumble bees generally favor those that are flat, with vertical symmetry. Also, movement of the flowers by wind is, we have seen, a strong source of attraction.

First discovered by observers at the beginning of the 18th century, visual landmarks present in certain flowers have recently received more scientific study. Flowers have groups of lines, dots, and colored spots visible or invisible (except in ultra violet light) usually at the base of the petals. These are guides to direct bees toward hidden nectaries. Olfactive guides compliment, at short distances, the optical ones and help insects to perceive the shape of objects (Meeuse, 1969). Even more extensive research has been done on the subject. It has been shown that nocturnal white flowers do not possess a visible nectar guide. Magnolia flowers which are normally pollinated by beetles, do not have any guides. Hymenoptera which land on these flowers generally go back with an empty pollen basket because they do not succeed in finding the center of the flower. In addition to colored or odoriferous nectar guides, there are also guides composed of hairs, fringes, and gutters. When touched by an insect, they help direct it to pollen.

The odor of flowers attracts pollinators. Some are sweet and pleasant to humans, others, frankly, unpleasant, as those of species of *Amorphophallus* or *Rafflesia*, giant flowers of the tropics. *Amorphophallus* spp. (fig. 21) smell of rotten fish and molasses, and attract beetles in large numbers until pollination is achieved. African species of *Staplia* attract flies by emitting odor and mimicking the color of rotten flesh. So close is the resemblance that the flies lay eggs inside the flowers. Even the hairs of the flowers resemble mold growing on rotten meat. *Rafflesia arnoldi*, an Indo-Malaysian plant, is exclusively pollinated by flies which are attracted by its color and odor. Blow flies are attracted to yellow or orange flowers even if they have no smell.

Generally, the smell of flowers is agreeable to both humans and insects, but these odors are not mechanically continuous. They can be discharged when the plant is shaken by an insect alighting, or by the wind: they can be emitted at certain fixed times, perhaps only in the evening, or throughout the night. The odor ceases after pollination is completed, or when the flower withers. Pollinators, whether insects, bats, or others, must be adapted to these rhythms. Birds, always diurnal pollinators, are attracted by bright colors instead of odor. Nectar production too follows certain rhythms. There exists a synchronism between the activity of the pollinators and the secretion of nectar.

It is very difficult to classify the odor of flowers. Van der Pijl (in Faegri and Van der Pijl, 1966) distinguishes the absolute odors (particularly smells from plants such as violet, rose, geranium, jasmine, and so on) and the imitative odors, *i.e.*, the odors which "imitate" ones to which the insects are conditioned by instinct and experience (sexual odors from species of *Ophrys*, odors of rotten meat, excreta, decayed fruit, and others). *Arum conophalloides*, for instance, imitates the odor of mammal skin. These plants attract small bloodsucking Diptera (Ceratopogonidae, Simuliidae, and Phlebotomidae). Mycetophilid flies are attracted by the fungus smell of certain inflorescences. Certain odors seem to provoke mating between insects, namely odors from Umbelliferae.

The question of odors which repel, or of marker odors left by Hymenoptera, is still being discussed. Also being debated is Kullenberg's hypothesis, according to which, the smell of some flowers imitate the body odor of the pollinators themselves.

In some way or another, insects reach the entomogamous flower's nectar, pollen, nutritive tissues (hairs of certain orchids, aroid tissues), waxes, resins (orchids, Euphorbiaceae), and aromatic or nutritive oils. Some find a breeding place in the flower, and probably often utilize other features, but they never use all of these elements at one time.

Nectar is composed of carbohydrates, but some flowers, the poppy for instance, have no nectar. Instead, they offer the pollinator a quantity of pollen rich in proteins. The insect is attracted by the smell of the flower, its color, and sometimes the flower is warmer inside. This feature is rarely reported, but many flowers use this combined with a fetid odor. Some even temporarily capture their visitors, sometimes holding them for the entire night, time necessary to complete pollination. The temperature of the spadix of *Arum orientale* does not go below 43° C, even when the outside temperature is as low as 15° C. A temperature higher than the ambient is often produced by arctic flowers in order to attract mosquitoes.

Orchid flowers of the genus *Catasetum* (Catasetineae) are visited only by male euglossine bees, even though they do not produce nectar, and little pollen. Gongorinae, and some other orchids, are likewise pollinated only by males of these bees. The reason for this has been discovered only recently. Odor attracts these males which collect aromatic oil secreted by a special tissue of the flowers. This is stored by the bees in a specially modified groove of spongious tissue of a chitinous nature on their hind tibiae. Often host specificity in tropical America is between an orchid species, its odor, and euglossine bees which visit it. The use of this perfume is doubtful, but it seems that it may be a way of marking the bee's territory during its nuptial parade. This behavior may be compared to the ritual of color display among birds of paradise and some other groups when vagabond males assemble to attract females. However, odor replaces colors for these bees. A similar case with *Gongora* spp. is considered later. This phenomenon is termed gamokinesis, or social pollination.

Females of anthophorine bees in the same manner collect a nutritive oil from numerous plants, stock it on the hairs of the hind legs, and feed their larvae by mixing it with some pollen.

Orchids of the genus *Maxillaria* also furnish a false pollen which clings to the flower labellum (or lower lip). Other plants produce waxes or resins in flowers for the same purpose.

Certain plants have transparent, windowlike spots on their petals near the stigma. Insects are attracted towards this false exit, but since these windows are located near the stigma they are drawn toward it so that pollination can take place.

Depending on the type of pollinating insects, flowers have either a landing platform (the orchids labellum which is the inferior lobe of the corolla) or nothing at all, if the insect hovers above the flower. Certain flowers "inform" their visitors that pollination has been completed: by withering, by closing, as in the case of peyote (a Mexican Cactus), by changing color for example the magnolia, or by ceasing production of the attracting odor. The African climbing lily, *Gloriosa superba,* changes its color from yellow to orange after fertilization and stops producing nectar. Beautiful and fragrant species of *Brunfelsia,* those strange American solanaceous plants, also share this characteristic. Their old white flowers coexist with the recent purple ones on the same plant. Many Borraginaceae also change color the same way. A red spot replaces the yellow nectar guide in the old flowers of horse chestnut trees and the effect is to "ward off" bees and direct them toward young flowers.

Many insects are in the habit of "stealing" nectar, without pollinating, by piercing holes in the base of the flower. Bees doing this often rob pollen. Some anemophilous flowers, such as plantain, are regularly pollinated by syrphids (Diptera). Sometimes they lay eggs in the flower and their maggots feed on the flower's pollen. This is true of cocoa, elephant's ear plant, and breadfruit, plants whose flowers and fruit feed their pollinators on the spot.

Some countries are devoid of nectariferous plants in any quantity. Because of this, a species of *Xylocopa* bee, usually the pollinators of *Calotropis procera,* which has a very short flowering season, must revert to other flowers for food. Or, during the dry winter of Cape Verde Islands, adult *Xylocopa* spp. pierce the base of the conical flower of *Tecoma stans* (Bignoniaceae) to get the nectar since it cannot penetrate the yellow coneshaped flowers, which are too narrow for its big body.

Some recent discoveries have proven the existence of amino acids in certain nectars. The concentration of this seems greater among more evolved plant families and especially those pollinated by Lepidoptera (except sphingids) or the ones that smell like carrion. Bees which receive their proteins from pollen generally visit flowers with glucose nectar. That plants supply proteins to insects via nectar is a recent hypothesis. Myrmecophilous plants are known to produce this kind of protein for ants.

According to Baker (1973), it seems that the chemical composition of nectar varies according to the plant and its pollinators. In addition to sugars and amino acids, some lipids, antioxidizers, alkaloids, and proteins are produced. Amino acids seem to be used when nectar is the only food source, as among adult butterflies or, for instance, when it is needed to attract blow flies. hummingbirds and hawk moths (sphingids), as well as bees and flies, seem to particularly benefit from lipids.

Since bees receive their proteins from flower pollen and birds obtain theirs from insects, they eat from inside the flowers. The flowers visited by these two groups are relatively poor in amino acids.

A correspondence between the distance of the nectar inside the flower and the length of mouthparts of the corresponding insect certainly exists. It is often a lock and key relationship. This is a strong indicator of coevolution between plants and insects. For instance, the relationship between the Malagasy orchid, *Angraecum sequipedale,* which is provided with a nectar tube measuring 35 cm, and a sphingid moth with a long proboscis, has been demonstrated. The same relationship has finally been found in tropical America between the orchid *Habenaria* sp. and "its" moth with a long proboscis used for sucking nectar. However, the relationships between some flowers and their pollinating insects are still to be determined, if, as is very probable, they do exist.

Almost all the flowers of milkweeds (Asclepiadaceae) are able to close and hold a visiting insect inside for a time before releasing it. During the "incarceration" of the insect, the flower's pollen is released on the visitor in the form of pollinia and pollination is completed. As we have seen, pollinia are sacks of pollen with an adhesive stalk. These sometimes resemble the palpi of the visiting insect, especially when they are attached near the mouthparts. Some flowers have inwardly directed hairs which have an eelpot effect on foraging Diptera, for example, *Ceropegia* spp., looking for nectar, or temporarily trapping *Xylocopa* spp. (as in *Calotropis* flowers). Pollination of *Calotropis procera* and the attaching of pollinia on *Xylocopa* legs has been studied in Dakar by Jaeger (1971). This also occurs in Vietnam and Thailand, either with this same species, or with *Calotropis gigantea* (fig. 22). In Thailand, this last plant is also fertilized by leaf- and flower-eating species of Chrysomelidae belonging to the eumolpine genus *Platycorinus.* In Senegal, fertilization of the flowers takes place during the dry season (January) when the pollinia are mature, but frequently I have observed some species of *Xylocopa* collecting nectar long before the pollen is ripe. Other Hymenoptera, including bees, and some Lepidoptera, also are involved. The flower may be permanently inhabited by small ants and Thysanoptera which do not seem to play an important role in fertilization. It is when moving and trying to disengage their legs from the interstaminal groove that *Xylocopa* bees pull out the retinaculum, a small glandular mass, and its pair of pollinia. Species of *Xylocopa* pollinate many other flowers including flowers of species of *Acacia* when species of *Calotropis* are not flowering. These bees are not specific on calotropis flowers unlike the specificity of certain insects to certain orchid flowers.

FIG. 22. *Calotropis gigantea* from Thailand. These plants are pollinated by various species of Hymenoptera, including species of *Xylocopa* which are kept temporarily inside the flower as prisoners. The plant is eaten by a danaid butterfly, some moths, and various beetles, species of *Platycorinus*. The insects which live on *C. gigantea*, including yellow aphids and locusts, show aposematic colors because the toxic properties of the plant are transmitted to the animal hosts.

As with birds and bats, there must be a correspondence between the distribution of insects and the flowers to be pollinated. In species of the genus *Aconitum*, for instance, pollination is effected by bumble bees with long probosces but the species do not have overlapping distribution. Some species of *Xylocopa* replace bumble bees in tropical Africa, even though these insects coexist with the bumble bees in temperate areas.

Another case of the adaptation of bumble bees to the life cycles of plants is exemplified by a composite from Himalaya, *Saussurea sacra*. This plant is covered with hairs so dense that it resembles balls of fur. It is often covered with snow at these high altitudes. At the top of the plant is an opening through which the bumble bees enter. They often spend the night inside the flower where it is safe and warm and, at the same time, they pollinate the plant. The flowers of this plant are odoriferous, which is rare for the flowers of Compositae, but this seems to be a relationship with its entomophilous pollination. The remarkable species of *Saussurea*, all extremely woolly, are represented by a score of species growing above 4,000 m and often as high as 5,500 m in the Himalayas. Other species live in European mountains. *S. alpina* is sweet scented and must have a similar life cycle.

Less useful as pollinators than bumble bees, honey bees, which, even though they may succeed in penetrating the flowers of red hot poker, *Kniphofia* spp., in the high plains of Ethiopia or East Africa, they cannot get out. The adaptation for flower pollinators in this case has never been completed. Why? It would be interesting to study these plants in their natural surroundings in an attempt to find out more about these relationships.

Each species of bumble bee (*Bombus* spp.) has its preferred flower, but occasionally they will also collect pollen from other flowers. The fact that they are able to use other flowers that exist in the same area, in case of rapid change in the composition of the vegetation, is a remarkable adaptation. Thus, these polyphagous species often have an advantage over monophagous species with their obvious lack of plasticity.

The attraction of the enormous morphological and esthetic diversity of the orchids hides a sexual life so elaborate that it might be compared with a living "Kuma Sutra." This sophistication depends on the very important role played by many species of insects. About 35,000 are known to be adapted in a unique way to species of orchids, and sometimes even snails are needed to pollinate some of the rare species. Certain orchids have no smell;

others produce a strong odor to attract the insects, each varying according to the request of the "customer." For instance, some *Dendrobium* spp. change their perfume from that of a lily-of-the-valley during the day to that of a rose during the night or from that of a helitrope in the morning to lilac during the night. Some orchids produce an odor similar to carrion, frankly unpleasant. They keep their flowers intact for months waiting for pollination, a feature useful to greenhouse orchid growers. Once pollinated, some orchids stop production of their attractant smell, while others lose their color and rapidly wither.

We actually know that orchids are a family in rapid evolutionary development as is shown by the many hybrids between species, or even genera, produced by growers. So far about fifty different specific odoriferous compounds have been isolated in orchids. When, as often is done, they are subtlety mixed, this maintains the attraction of "their" specific pollinating insect, and prevents natural cross-pollination.

Some orchids produce a rotten meat odor and their flowers even imitate the color of rotting flesh. These orchids are fertilized by flies which either lick an agreeable secretion produced by the plant, or they are attracted by this deceptive odor, since the flower has nothing to offer other than this illusion.

However, this fraud is not the rule and many orchids offer a drink of nectar and food on hairlike projections to the insects which pollinate them. The food is rich in proteins and is eagerly chewed by the insects.

Certain orchideating chrysomelid beetles, *Petauristes* spp. for example, feed on parts of the flower, and at the same time fertilize it, as is done by species of *Aulacophora* on the flowers of melons and pumpkins (Curcurbitaceae).

Orchids also attract insects by their color, their shape, their pilose fringes, their trembling hairs, and many other "artificial" means. Insect guides are either morphological, or nectarigenous areas and are a trap used to attract insects as they collect pollen or drop it on the stigma. Generally orchid flowers are hermaphroditic, but autogamy sometimes occurs. If the sexes are on separate flowers, these are dissimilar and sometimes have been described as separate species with different names. In the latter case, the plants are dioecious, but there are exceptions as some are known to be diclinous monoecious species (different flowers on the same plant). As previously mentioned, orchid adaptations to attract insects are very numerous. Darwin was the first to outline these peculiarities. Since Darwin, many additional facts have been, and continue to be, discovered. Mainly, it is now known that very few species secrete "free" nectar. Almost always, the insect must find its way to the nectariferous tissue, although sometimes they may pierce the flower to do this.

The rostrellum of the stigma, or several structures covering it, often form obstacles to prevent self-fertilization, and often the pollen sacs are an agglomeration of ovoid masses forming pollinia which, as we have seen, are taken away to other flowers by the insect visitors. When they reach another flower these bend, in about 30 seconds, and meet the receiving surfaces where the pollen grains are deposited. As with rhododendrons, white orchids have a strong odor and attract moths at night. The flowers produce pollinia which attach to the head of the moth, sometimes even to its eyes.

The structures of the flowers of species of *Pterostylis* remind me of fly-catching plants. Insects land on the flower's labellum, a large petal which provides the landing platform in front of the flower. This is hinged and it closes against the flower to force the pollinator to follow the path to the inside. In half an hour the labellum opens again. Many other structures which remind us of the temporary trapping of *Xylocopa* species by the *Calotropis* sp. flowers exist among some orchids. Sometimes an insect is projected to the inside of the flower by a shaking of the labellum, or by rapid movement. The hinge system is sometimes even more complex and actually carries the insect to the nectar, its reward, and to the pollinia, its ransom.

Species of *Gongora,* orchids from tropical America, attract bees with odor from a scent gland, the osmophora. The slippery surface of the flower resembles a toboggan and causes the bee to slip on its back into the flower where it attracts the pollinia and fertilizes the stigma by the pollinia of another flower (fig. 23). Other orchids have different ways of capturing bees including the transparent windows, mentioned previously, which guide insects toward pollen or toward the stigma before they are released.

Other solutions are mindful of the tactics used by carnivorous plants. Some insects literally become intoxicated by the odor of a smelly tissue; then they fall down into the liquid collected in a pailshaped labellum, but they are able to escape through a gutter which directs them to the pollen or to the stigma. A similar process is found in a Central American orchid. The first hymenopteran to visit has difficulties coping with the rostrellum, but the later ones to arrive easily escape, soaked, drunk, and often having completed their pollinating function. It has been said that this orchid stops producing its odor if the insect takes too long to escape, but starts it again the next day, probably method of avoiding self-pollination.

Some reports indicate that certain orchids, such as species of *Gongora, Stanhopea, Catasetum,* and others, generally cause intoxication of insect visitors, and that the anthropomorphic explanation given is that it "calms down" the hymenopteran. The intoxicating substances, retained on the legs of the hymenopteran, might also play a role as markers or as a sexual attractant.

Other orchids, among them, species of *Catasetum* growing in tropical America, are provided with triggers (fig. 24), floral appendices with a sticky viscidium in front, which release a catapult and attach the pollen mass to the insect. They are capable of projecting pollen a distance of one meter. Chiefly flower eating insects visit *Catasetum* spp. They go directly to the pollen mass, although euglossine bees also are attracted.

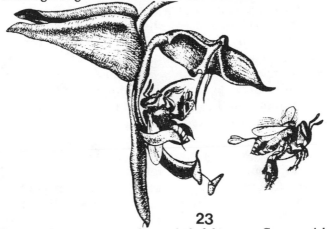

23

FIG. 23. Pollination system of an orchid of the genus *Gongora*. A bee which penetrates into the flower meets a gliding surface and falls tobogganlike along the column (gynostene), where it strikes the visoidium (retinacle) which adheres to the insect's abdomen. The bee flies away and takes with her the pollinia (after Arditti, 1966).

Insect traps occur in species of *Arum*; others among species of *Pterostylis* and *Masdevallia*, plants with pitchers and a cover composed of a labellum which can stay closed for as long as 30 minutes.

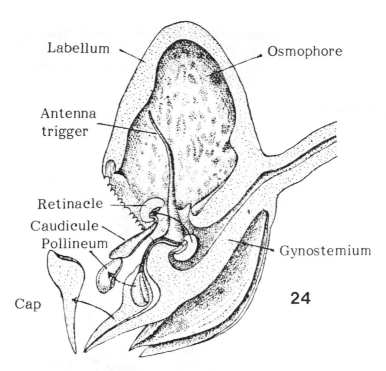

FIG. 24. Trigger mechanism of an orchid of the genus *Catasetum* which provokes pollination once released (after Arditti, 1966).

Certain orchids without nectar, oil or edible tissue become prostitutes (a fact known since at least 1916, and restudied by Kullenberg [1961] and others since). By mimicking a female insect, the flowers are fertilized by various Hymenoptera which practically are "masturbating" on the female insectlike flower. This is what Jaeger (in Grasse, 1976) calls "flower sexual partners." This strange similarity between flower and insect was noted as early as 1831, but this fact was not yet connected to the role in fertilization. As stated by Huxley (1974) the flower offers its own flesh to satisfy her sexual desires.

Species of *Ophrys* orchids in Mediterranean Europe (fig. 25) are distinguished by the form of their labellum. They resemble in shape, color, and velvety structure, the body of an insect. Australian species *of Cryptostylis,* the South American species of *Paragymnomma,* and several other orchids also mimic insects to attract them. Numerous other examples certainly may be found wherever orchids grow. These orchids, devoid of the usual attractants, are, however, visited by Hymenoptera, flies, and other insects which pollinate them. The American orchid, *Trichoceros antennifora* has evolved a similar relationship, not with an hymenopteran, but with certain flies.

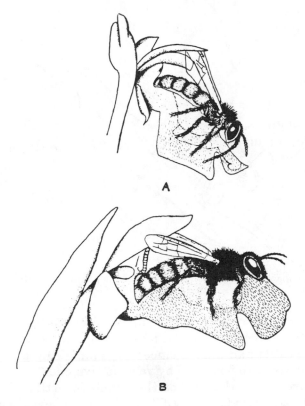

FIG. 25. Mating reactions between: a. *Andrena maculipes* (the bee) and *Ophrys lutea* (an orchid). b. *Andrena mactae* (the bee), and *Ophrys fusca* (an orchid) (After Kullenberg, 1961).

Ophrys apifera, an English species, self-pollinates because it has lost its adopted insect pollinator as its sexual partner. Mating between insect and flower take places only when the males are newly hatched and the females are still very rare. As we have already seen, the privilege of fertilizing the orchid by male insects exists elsewhere in other orchids and for other reasons.

In North Africa, it is *Scolia ciliata,* a bee, which pollinates *Ophrys speculum* during false copulation. In Northern Europe,

Ophrys muscifera is pollinated by two species of wasp of the genus *Gorytes*. Numerous similar cases in the Mediterranean area involve other species of *Ophrys*, and *Scolia* bees. Others are pollinated by species of *Eucera* and *Andrena* bees. Kullenberg (1961) has observed bigeneric attractiveness of *Ophrys scolopax* in Lebanon to males of species of *Eucera* and *Andrena* bees. To facilitate mating, the walls of epidermal cells of the labellum of *Ophrys* spp. is relatively strong. The primary stimulation is essentially olfactive, followed by a tactile and visual stimulation, the last is secondary and depends first on tactile stimulation.

A layman imbued with a dose of finalism, sees that the flower morphology of *Ophrys* spp. resembles an insect somewhat, mostly Hymenoptera. These "insects" have false eyes, antennae, wings, and the "body" hairs are similar in shape, color, and texture to the hairs on the female insect. This is not to imitate Bernardin de Saint Pierre, and his extreme finalism, but rather to suggest that there is at least a partial truth in the observation.

The true attraction for insects, however, comes from an odor similar to that produced by the female insect, a true pheromone mimic secreted by the plant itself. So strong is this smell that males try to find flowers when the material is impregnated on paper, and at least the males of one species prefer the orchids to their own females. Flower morphology has only a secondary role in attracting the insect, if such a role actually exists. To compensate for its immobility, the labellum on which the insect lands is dark purple, or brownish with clear markings, or is a mirror spot which reflects ultra violet light. Once landed on the orchid labellum, the insect finds the curved hairs. These "persuade" him that he has really found the female and he starts a long and vigorous false copulation. The copulatory organ is extruded and is inserted into velvety areas of the labellum. Males, excited by smell, sometimes fight for the possession of a flower--a poetic picture isn't it?

However, the orchid is not enough for its loving partner and although the penis is erect and rubbed without interruption, there is no ejaculation, at least for those visiting species of *Ophrys*. Sperm, however, has been produced when copulating with species of *Cryptostylis*, an Australian orchid, pollinated by the ichneumon wasp, *Lissopimpla semipunctata*.

As A. Huxley (1974) wrote with humor, "even so, the insect flies full of hope from flower to flower, practices again a simulacrum of copulation, and we hope finds its own pleasure, but, with great efficiency, it carries away pollen." Sometimes the labellum is violently bitten during these actions.

These prostitute orchids vary greatly in shape and pilosity. This agrees with their "partner," their geographical origin, and the species of orchid. *Ophrys lutea,* for example, causes inverted mating because of its floral structure. Anyway, this resemblance between a flower and an insect is sufficient. It lacks any other source of attraction, and since the insect does not get any remuneration for the act of pollination, it is in all ways remarkable. It must be the fruit of a long evolution which began with a nectariferous plant. For these orchids, the erotic solution is their unique chance of survival, since cross-pollination is for them a necessity.

Orchids are not the only example of production of a pheromone mimic. Truffles, subterranean fungus growing several inches to one meter below the soil surface, can be detected by its aroma which resembles an androsterone of boars as well as of the human male. It is difficult to prove, but apparently this stimulant plays no role as a human sexual pheromone, but for the sow it certainly does. In the south of France female pigs have been trained to locate truffles by scent. If the sow eagerly digs in soil for a truffle, it is reacting to this scent by a mating behavior. Why we ask? Perhaps, coincidence, but also perhaps the fungus has "evolved" this solution in order to be disseminated. From wasps to pigs, there is a big distance, but in both cases the plant has developed animal mimicking pheromones to effect its reproduction. There may be many other still unknown cases.

Male tropical bees of the genus *Centris* are attracted by the vibration of orchids of the genus *Oncidium,* and believing they see a rival visiting their territory, charge the flower and thus get the pollinia on their heads. In this case, neither odor, nor food, nor color, plays any part in the attraction. It is only the territoriality of the insect that causes this behavior. How did such parallel evolution (or coevolution) occur? It is rather difficult to imagine.

Among the scolicid wasps of the genus *Campsomeris,* the females take the labellum of species of *Brassia and Calochilus* orchids to be a predatory insect. Such is at least one interpretation given for their behavior. They attack that false intruder, sting it, and hence pollinate the plant. This is what Van der Pijl and Dobson (in Gilbert and Raven, 1975) named pseudoparasitism (we could also say pseudopredation). Also, with this deception, the insect uses its energy, free of charge, in the service of the plant. The exchange is decidedly out of balance.

Still other insects are imitated by orchids. Species of *Oberonia* in Sri Lanka produce a flat rachis with small greenish flowers imitating, to human eyes, honeydew producing aphids. Is this the reason for the pollination of those flowers? These facts have to be studied further.

Saprophily, or egglaying by flies in the flowers imitating the smell and color of rotten flesh, is not always a deception for the insect. In several cases, the larvae do not die but develop normally, producing a new generation of pollinators.

However, Dobson (in Hocking and Raven, 1975) writes that most of the relationships between insects and orchids are based on deceptions for the pollinators. In many areas, flowers are separated by time and space and no energetic equilibrium between pollination and pollinated can be maintained. There is no symbiosis in the strict sense, but instead, parasitism of the insect by the plant. Pseudocopulation, pseudopredation, pseudoparasitism (for the parasitoids), pseudoterritoriality, or pseudoantagonism, pseudopollen on the labellum, pseudonectaries, pseudostamens, pseudoperfume, convergence in shape and color with the neighboring plants, and abundance of nectar, are frequent phenomena among orchids, all of which are becoming better known, and utilize, in some way, the insect's appetite. In more than half of the known orchids they do not provide food for their pollinators, but, at the cost of a rather small energy expenditure, provide them only with illusions. Other species, apart from nectar, provide pollen, edible pseudopollen or other food bodies. The edible pseudopollen is basically made of starch as in species of *Maxillaria,* but more often they are empty and not edible as in species of *Polystachya.* Thus it is not only species of *Ophrys,* with prostitute flowers, that are selling illusions.

Two sympatric species of *Oncidium,* cited previously, exist in the Bahamas. These are infertile between themselves, but are exclusively fertilized by the males of *Centris versicolor* (because of pseudoterritory inducement), in one species and by the females of the hymenopteran in the other. Also in the latter, these orchids are mimics of species of *Malpighia,* which are themselves nectar producing. Hymenoptera are the dupes, but even so, these complex associations have evolved and are maintained. The damage caused to *Centris* species must be slight, since it never ceases.

Pollination, with a long evolution of more than 200 million years, became a problem for the flowering plants only when they started to develop away from water. The ones which return to water, such as *Zosterae*, have found new solutions completely independent of arthropods. Even though terrestrial plants, the male gametes of species of *Cycas* and *Ginkgos* are antherozoa. These are mobile and swim for weeks in the moisture of their environment before reaching the female nucleus. For higher plants, the microsporophyte produces pollen which, thanks to the development of pollen tubes, joins and fuses with the ovule to become the egg. The pollen of the zoogamous flowers is sticky and its surface is covered with spines, crests, and rounded points, unlike anemogamous pollen which is smooth, light, dry, and powdery. Anemogamous flowers are small, greenish, without any bright colors or nectar. The energy cost is less; the wind works free of charge.

A flower with a deep corolla cannot survive if there are no insects with long probosces, or birds with a long beak that can suck out nectar and pollinate it. The contrary is not evident since insects or birds with a long appendage can pollinate very well a short corolla plant. However, flowers and pollinators seem to evolve in parallel. It has even been said that insects, and perhaps mostly birds, have been the creators of the evolution of the flowers in the primitive tropical forest.

Most of the pollinating Hymenoptera have been mentioned previously and their model, so very well known, is the honey bee. However, numerous dipterans and a few beetles, also have long sucking probosces. Many beetles, however, are rough pollinators of primitive flowers, such as species of magnolias and Calycanthaceae. These last plants do not provide nectar or pollen, but food made of modified pseudostamens.

Many insects, especially beetles, and some birds rob, as we have noted previously, nectar and pollen from the flower by piercing the corolla, and hence, they do not pollinate. Apparently this was once true of most flowers, but now it is chiefly the very complicated flowers that are victims of this predation. It is now thought that angiosperm evolution has been rapid because of this. As early as the Carboniferous period, insects must have fed on the spores of vascular cryptogams, followed by pollen of gymnosperms as early as the end of the Jurassic. Later, by selection, flowering plants produced nectar, odor, colors, guides, and other modifications until they reached the stage of "one insect for one flower" which, reducing possibilities of pollination, provokes a highly specialized adaptation of floral structure. Energy balances, or pollination costs, have been calculated by several authors in calories provided by nectar sugar of the flowers visited. Insects must rapidly visit the flower to procure an energy profit. Many flowers must, therefore, appear at the same time, or grow in colonies, to reduce to a minimum the distance traveled by the pollinators. These calculations are, however, hypothetical, and no general statement applies to all pollinators. For instance, butterflies are less in need of energy than bees. Butterflies do not feed their larvae; bees do.

Hocking (in Gilbert and Raven, 1975) calculated the minimum distance covered by a nectariphagous female mosquito in the Canadian tundra. He estimated that sugar contained in the ventral diverticulum of a female mosquito's gut gives her a flight range of 25 km and that a single catkin of *Salix arctophila* produced enough energy for 950 mosquitoes/km per day. Further, it is known that flowers pollinated at low temperatures produce more calories (in quantity of nectar) than those which blossom at high temperatures. These flowers often grow in patches, a valuable asset in the Arctic and at high altitudes. The more the plants are grouped together, the shorter the distance to be covered by the pollinator and smaller the energy expenditure. This is probably a genetic differentiation characteristic in local populations of plants.

There is certainly an equilibrium between plants and insects, writes Mesquin (1971), because in temperate regions, insects compete for flowers, but elsewhere it seems to be the plants that compete for pollinators. To do this they must be "rivals" in developing special attractants. In American, African, and Papua New Guinean tropical mountains, birds are more efficient pollinators than insects since they pollinate even during periods of cool, rainy weather, common in tropical mountains. This is something that insects cannot do. This fact is well known for the species of *Lobelia* that depend on sunbirds (nectarinians) in such places as Ruwenzori or Kenya.

Dealing with energy balance, there is the well known example of the beautiful species of *Heliconius,* tropical and subtropical butterflies. These pollinate species of *Anguria* (Curcubitaceae), but their caterpillars feed exclusively on Passifloraceae. Adults live for six months, an exceptionally long and active life for a butterfly. This is attributed to the fact that the butterfly feeds, not only on nectar, but also on pollen rich in proteins. However, exclusive pollen feeding does not lengthen the very short lived adult moths of the moth family, Micropterygidae.

Just as we have classified the phytophagous insects into categories, monophagous, oligophagous, and polyphagous, Hymenoptera, including bees, have been divided into monotropic (flower visitors), monolectic (pollen gathers), oligotropic or oligolectic, and polytropic and polylectic species. However, most of the Hymenoptera have the tendency to be generalists or no more than moderate specialists, and only in certain genera (such as species of *Apis, Halictus, and Megachile)*, is the percent of pure pollen gatherers very high.

As the plant is able to induce the speciation of phytophagous insects by selection to form biological races, the pollinators, such as bees, can accelerate speciation of "their" plants when sympatric races exist.

The flower being the seat of reproduction, it is normal that the plant develops every way it can to protect it. It is the flower of *Cannabis sativa* that has the highest concentration of the drug marijuana, or hypericin in the flowers of St. John's wort, pyrethrum in the flowers of *Chrysanthemum,* and so on. The flower is also morphologically protected, from intruders, by various structures in a multitude of shapes and strengths which attract only the most common and hence, the most useful pollinators. Their survival and multiplication is at that price. Mammalian and avian pollinators most likely preceded the more refined bee pollinators. Initially, flowers, or analogous structures, must have been robust, with protected ovules, mainly because beetles were probably the first important pollinators, preceding many other insects (Hymenoptera, Lepidoptera, and Diptera) which treat flowers with much more care. If we believe certain authors, the primitive "flowers" of the Caytoniales were pollinated by small reptiles during the Mesozoic and fossilized excreta has been found with the flower's print.

In nature, pollinators vary according to climatic areas. For instance, Lepidoptera tend to replace bees in temperate mountains; flies dominate in arctic areas, ants in semideserts, birds in tropical mountains, and, where specially evolved pollinators are missing, "primitive" ones, such as beetles, get their chance. Among large insects, only adults pollinate flowers although certain small insect larvae can pollinate, but their role, in general, is negligible.

It is possible that entomophily is primitive among the angiosperms and that anemophily is a secondary phenomenon, but the question is complex and far from being resolved. When some flowers are introduced into a new geographical area, pollination is usually rapidly accomplished by insects or birds, but they are never as efficient as those in the original country of the plant. This can only be called preadaptation instead of coevolution.

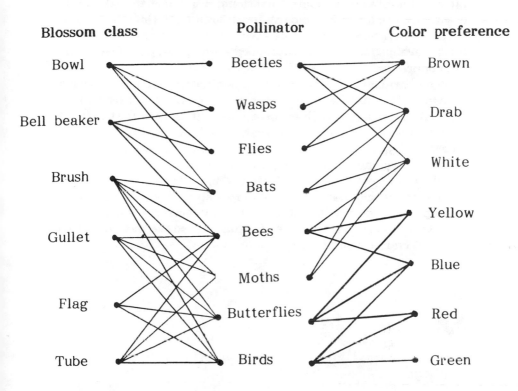

FIG. 26. Interrelationships (harmony) between pollinators and various flower types (after Faegri and Van de Pijl, 1976).

In any case, when cross pollination is obligatory, this contributes to gene flow throughout the population, mostly thanks to insects. This gene flow certainly contributes to the variation and diversification of the species and varieties within a plant genus. It is, for instance, almost certain that some pollinating Lepidoptera and Diptera show strong preferences for a color variation (morph) of a given flower among polymorphic species. We have seen previously the example illustrated by species of bumble bees. The variations, shown by the plant species of *Lantana*, *Raphanus*, *Cirsium*, and many others, are well known. For example, in species of *Cirsium* with white or red flowers, it is the white morphs that are preferred by their pollinators. The abundance of white forms is interpreted as an adaptive answer from the plant to conditions of low pollination. Without denying that ecological factors can influence certain variations, it is a fact that the flowers often change color in mountainous areas. On the other hand, pale or white flowers are more attractive during the night or at dusk, perhaps because they are more visible, and therefore, their abundance in a dense forest is explained.

The harmonic relations between pollinators and blossoms are shown in fig. 26 where the blossoms are classified by shape. Although a variety of pollinators are attracted, note that there is remarkable selection both as to shape of the flowers and their color. No doubt, this strongly indicates selection and adaptation between the plants and the animals involved.

We have seen that coevolution of insects and plants, even though the mechanism is not yet very well understood, is due to phytophagous insects, to pollinators, both vertebrates and invertebrates, and it is certainly insects that impose the most selective pressure on these plants.

"Water flows by
Time goes by, flowers disperse
Who would like to hear it.
It I told to wait."

-Japanese poem, XIst century

Chapter 13

Entomochorous Plants

Without the risk of being accused of anthropomorphism, it can be said that altruism does not exist in the plant kingdom. We have seen previously that when a flower gives up its nectar this usually ensures its pollination; when myrmecophilous or carnivorous plants provide food for their guests they get in exchange, first protection, and second, the nitrogen they need. When fruits are attractive to animals, seed dispersal is assured. We will review in this chapter insect dispersal of plant seeds. This is termed entomochory.

Entomochory is mostly myrmecochory, *i.e.*, transportation of seeds by ants, but other insects are sometimes dispersal agents. It is evident that during their travels, many animals disperse seeds from their body fur, or in their excreta. This includes dispersal by vertebrates, insects, crustacea, and even earthworms. A great part of the animal world contributes to this dissemination. During the Mesozoic era, some reptiles were herbivorous, and they must have been responsible for the dissemination of gymnosperm seeds including those of the *Ginkgo*, and later those of angiosperms. Some of these seeds have been found in the imprint of fossil stomachs. Now the role of reptiles is greatly reduced or nil. During the entire timespan, the role of insects, in seed dispersal, must have played an important part, developing after the appearance of angiosperms during the Cretaceous period, *i.e.*, at the end of the Mesozoic.

Insect dissemination of seeds and spores of plants takes place in three different ways: a) by swallowing them and voiding them undigested in their excreta (endozoochory); b) by taking them into their nests where they eat only oily body or caruncle adjacent to the hilum in certain seeds (elaiosoma); and c) by adhesion of spores or seeds to their body (epizoochory).

The distribution of the seeds by the first method is rare, but is a common way for spore dispersal. However, Darwin, in the "Origin of Species," mentions that in Nepal he reared seven different plants from the excreta of several Napalese migratory locusts. The spores of specific fungi depend on insects and even slugs, for dispersal. This is the way *Phallus impudicus*, a fungus with a strong odor, found in temperate regions, and its tropical relatives, is dispersed. Flies attracted by the fungus' odor, eat the mucus spores, but these are undigested. Instead they are excreted elsewhere. In the excreta of one fly, specialists calculated that there were in it as many as 20 million spores.

Beetles, namely certain Chrysomelidae such as species of *Diabrotica* and *Chaetocnema,* play a modest role in seed dissemination, but they are responsible for the dissemination of spores and bacteria. Ridley (1930) mentions Trichoptera as seed disseminators of marsh plants, as well as other examples involving Orthoptera and Diptera. The latter two disseminate mosses *(Splachnum* and *Tetraplodon* species), and fungus spores. Collembola also carry lichen soredia. We could multiply the examples but the carriers of the greatest number of fungus spores are Diptera, Homoptera, and Hemiptera. They also disseminate bacteria. It is to be noted that rye ergot *(Claviceps purpurea),* a pest of many plants (ergotism or "*mal des ardents*"), is carried by flies attracted to its sugary secretion which is produced during a certain stage of the life cycle of the fungus as nectarinian spores. Thus nectar production is not a complete monopoly of the higher plants. In Europe, during the Middle Ages, the disease was widespread among the peasants and as recently as 1951 an epidemic of ergotism occurred in a small village in the south of France.

Termites play a modest role in the dissemination of seeds of flowering plants. They also carry spores and sclerotes of fungi which they grow in their nests. In Transvaal, the grass, *Cynodon incurvatus,* is carried in the seed and fragment stage into termite nests where it is stocked in the upper chambers of the nest. Large, nocturnal, termitophagous mammals, orycteropes, break open the termite nest allowing the seeds to germinate. The seeds in the giant flower of *Rafflesia* spp. are disseminated actively and passively by termites which invade their fruit and ovary. However, myrmechory is the mainstay in seed and fruit distribution by insects.

In temperate countries, myrmechorous plants are mostly herbaceous plants found in lower strata. Trees are anemochorous (seeds disseminated by wind) and berry bushes (in the arbustive strata) dominate. Their seeds are disseminated by birds, as are, for example, the epiphytic mistletoe (Loranthaceae) with red and white sticky berries. We will see later that this is not the same in the tropics. There many epiphytic myrmecophytes produce sweet and sticky berries, white or red (species of *Hydnophytum*, *Myrmecodia*, and related genera), which are disseminated by ants of the genus *Iridomyrmex* and not by birds. These ants are so aggressive against any vertebrate attempting to touch the plant, that we could hardly imagine that birds are able to reach the berries. Many myrmecophytes are myrmecochorous epiphytes. Let us cite as an example, the Asian species of *Dischidia*, which are also distributed by *Iridomyrmex*, but not in the same way. There are several reasons why ants carry seeds. First, they take the seeds of grasses to their nests and store them as food. Generally they eat the plumule and the radicule at time of germination, which, of course, kills the seed. However, many seeds are lost on the way and these eventually germinate. Some ants, such as *Formica rufa*, carry seeds to use in building their nest. The most favored seeds are ones with an elaiosome or oily body, or ones which exude oil to cover the test of the seed. Often ants drop seeds after having eaten this oily body. Seed distribution of aerial plants by ants is very efficient for more than just the myrmecophytes. Eleven different types of these seeds disseminated by ants have been described. The number of seeds carried by one ant alone can be considerable (almost 40,000 for only one worker of *Formica rufa* in 80 days). The distance traveled can be as far as 70 meters. Generally these are seeds rich in oil, such as those of *Viola*, *Veronica*, or species of *Corydalis*, *Cyclamen*, *Chelidonium*, and *Melampyrum* (Ridley, 1930). The seed caruncle, the elaiosoma, contains oil, proteins, and various fats.

A classic example is the large celandine, *Chelidonium majus*, a European plant common in hedges, garbage dumps, and on old walls, as cited by Schremmer (1976). Species of the ant genus *Myrmica* carry away these mature seeds which they drop on the ground or take to old walls. Only these ants are responsible for the establishment of the plant on these walls.

Myrmechory is extremely common in the tropics, where all strata of the forest are involved. The inevitably present Neotropical leafcutting *Atta* ants are kept in check by ants with aerial nests on myrmecophilous plants. These ants defend their habitat ferociously, as well as being responsible for seed dissemination of their plant hosts.

Harvester ants (subfamily Myrmicinae) live in arid and semiarid areas of the world. They collect seeds from their plants or directly from the ground under the plant. These they store for the dry season in special chambers in their underground nests. These so-called "biting" ants (because they have a stinger on their abdomen) feed on these seeds. However, if the seeds become wet, the ants take them outside the nests, spread them out to dry and then take them back again inside the nest. The Bible and the Talmud both mention these ants and their wisdom. It is evident that plants which germinate around those nests were not planted intentionally. They are only the seeds lost or rejected during the harvest by the ants.

It has been shown that seeds of violets cut into by the mandibles of ants germinate quicker than the ones not carried by them. Probably both the cutting and saliva together help in the germination process.

"A drop of amber from the weeping plant
Fell unexpected and embalmed an ant
The little insect we so much condemn
Is from a worthless ant become a gem."

-English translation of
Martial Epigrams.

Chapter 14

Coevolution

To determine the origin of food selection is difficult. It is evidently much older than Tertiary Baltic amber. For instance, hematophagy, or blood sucking among insects, appeared in very different groups and probably independently, as for instance, among fleas, the dipterous families Culicidae, Tabanidae, and Glossinidae, the bugs Cimicidae and Triatomidae, and some other families. Some groups considered to be purely phytophagous or aphagous, for example Chironomids, probably were derived from hematophagous ancestors, such as still exists in Australia and South Africa. To be hematophagous, insects should have had reptiles, small mammals, or at least batracians, to bite. The origin of this type of selection is very old and parasites such as fleas (from Diptera) or lice (from Hemiptera) have evolved much later along with their bird or mammal hosts. Lower Cretaceous fleas are known from Australia and the Cretaceous preSiphonoptera from Transbaikalia, described by Ponomarenko (1976), was probably sucking the blood of Pterosaurians. However, it seems that at the beginning, all of those hematophagous blood sucking insects were phytophagous or nectariphagous, as the males of mosquitoes still are. It should be noted that both male and female biting mosquitoes are also efficient pollinators, and in certain cases, specific pollinators in the Arctic regions for species of *Habernaria* (Orchidaceae). Biting bugs (Reduviidae) have certainly evolved by changing from sap suckers to blood suckers and rare ophthalmotrophic moths (moths that feed on blood from the eyes of cattle) come from similar fruit piercing species.

Phytophagy certainly started much before the appearance of flowering plants which date from the Cretaceous, or at least from the end of the Jurassic. It has been stated, for instance, that in the Gondwanian flora in South Africa, the leaves of species of *Glossopteris* were probably eaten by an insect during the Permian. In the Carboniferous, and probably much before, forest litter was occupied by detriticolous and phytophagous insects, and not only by primitive ones, such as Collembola and Thysanura, but also by the more advanced cockroaches (Dictyoptera). Aphids probably began by living on gymnosperms. Even the more specialized groups such as the primitive ancestors of the first Coleoptera probably date from the end of the Paleozoic. These attacked the available plants and, namely in the Permian, the vascular cryptogams, by then much diversified since their Carboniferous origin, followed by the gymnosperm Coniferae cycads, ginkgoes, and chiefly, the Bennetitales, a kind of prototype of the flowering plants. These, of course, are not the ancestors of the flowering plants. The sexual parts of these plants were enclosed by leaflike organs, almost floral in appearance, with their ovules protected by scales. This group resembled Ranunculacease very much and it is probable that these ancient false flowers were brightly colored, bisexual, and attractive to their early pollinators, not all of which were insects, however.

Coleoptera, such as species of Aulacoscelinae, primitive Chrysomelidae closely related to Sagrinae, still actually pollinate the American cycads, and other coleopterous families do the same for the cycads and the Gnetales *Welswitschia mirabilis* in South West Africa. When they are imported from the tropics, the cycad flowers are always frequented by beetles.

It was not until the Cretaceous that the association of flowering plants and insects evolved and developed as they are now. At the beginning of the Cenozoic, most of the present day adaptations must have existed if we take into account the state of evolution of the insects now preserved in amber.

The tendency of the flowers of the fossil Bennetitales to increase in size is very typical. They increased from about 5 mm to about 1012 cm during the Cretaceous. That was probably in relation to the appearance of large pollinating birds, and probably mammals. Insects must have played a role at the beginning of this evolution. Almost certainly specialized beetles such as the Chrysomelidae developed in the Cretaceous parallel to the development of the angiosperms, but this is not really known to be true until the Eocene. Although ants are known from the Cretaceous, it does not seem that bees appeared before the Oligocene, *i.e.*, just after the appearance of the true Lepidoptera. The most evolved of

all pollinators, the bees, appeared last, along with the plants
most adapted to that function, such as Compositae, Labiatae,
Scrophulariaceae, Papilionaceae, and orchids. Together, they
have achieved this coevolution, a result of the perfection of their
adaptations.

It also seems evident that before sucking nectar from flowering plants, lepidopterans ate only pollen. The present day Micropterygidae, one of the most primitive moth families, closely related
to Trichoptera, have chewing mouthparts used for this purpose.
In Europe, for instance, these insects visit buttercups (Ranunculaceae) and eat their pollen, which is toxic to other insects, the
bees among them. The mouthparts of the Micropterygidae are
used only for biting. Their long palpi gather pollen, scraping it
from the anthers and bringing it to their mouth where it is
crushed and swallowed. Theoretically, pollen eaters who previously could have been spore eaters, are polyphagous. It is
probably in this way that the Bennetitales were pollinated before
the appearance of nectar, though it is not sure at all that those
"flowers" did not produce some.

The nectariphagous insects, such as bees and other honey
gathering Hymenoptera, are theoretically polyphagous (with
certain restrictions as already mentioned) as are the adults of
Lepidoptera. They are not at all specifically associated with the
entomogamous flowers and some will collect, during the spring,
the pollen of anemogamous flowers, such as those of poplars,
Cyperaceae, and Gramineae. If beetles seem to have been the
first true pollinators, and still are often the pollinators of gymnosperms and primitive angiosperms such as Magnoliaceae and
Nymphaeaceae, Hymenoptera are generally associated with the
most highly evolved flowers, and probably the most recent ones to
appear, such as the orchids and Scrophulariaceae.

Leaf cutting and fungus growing ants in tropical America
show the unique peculiarity of being both monophagous for the
fungus they grow, and polyphagous, since they cut leaves of many
kinds of trees. This polyphagy is, however, only relative, and
varies according to whether the species attacks grasses, or trees,
or both. This selection, varying with a change of taste, helps to
protect the environment from total destruction of a particular
plant.

It seems that polyphagy is the primitive type. Monophagy and
oligophagy, including their botanical sense, developed gradually
by restrictive mutations and narrow adaptations to their food

plants. This could explain, for instance, why parasitic fungi and aphids develop two generations on two unrelated plant families. It seems, indeed, that if oligophagy (or monophagy) is primitive, and not secondary, fungi and aphids would have chosen related plants. It would be wrong, however, to consider polyphagous leaf miners as more primitive, since they belong to four different orders (Lepidoptera, Diptera, Hymenoptera, and Coleoptera) each highly evolved. Often polyphagous leaf miners, formerly monophagous, become polyphagous by secondary adaptation.

Finally, if primitive insects were polyphagous, a necessity during the Paleozoic and at the beginning of the Mesozoic when plants were only slightly modified, as certain reptiles, they must have eaten without much selectivity, vascular cryptogams, and gymnosperms. Some primitive plants probably defended themselves by concentrating metabolites toxic or repulsive to insects. The majority of insects avoided those plants, but some of them were able to accept them through selective mutations and preadaptation. These insects had an advantage over the others - the absence of competition. These toxic substances then became for these insects attractants or nutritional stimulants. What was probably at the beginning a chemical defense for the plant became the reason for the specialized associations between the insect and the plant. Such is the theory developed by Fraenkel (1959) and by others. The scheme is perhaps a bit simple, but the example of St. John's wort, toxic because of the secretion of hypericin and repellent to many animals, is a classic. The plant attracts, on the contrary, a whole group of chrysomelid beetles of the genus *Chrysolina* which learned to detoxify the poison. For these insects, the repulsive principle became attractive. Hypericin harboring plants are avoided by most of the large herbivorous mammals. They use knowledge instead of genetic programming as used by insects, for protection.

It is this parallel development of insects and plants, of biochemical barriers and the means to beat them, which has been called "coevolution" by Ehrlich and Raven (1964) as observed in the butterflies, but this is now applied to many associations between a host specific phytophagous insect and a plant. The chemical products responsible for this process have been termed "allelochemicals." These play a great role in the way an insect behaves, and also in host plant resistance.

Pollination by animals also represents a form of coevolution which has lasted for more than 200 million years. During the

Jurassic, some large dinosaurs fed on arborescent conifers. It seems that by the Cretaceous they moved over to angiosperms and then were poisoned by their alkaloids. This is a rather questionable theory, but it is certain that changes in fauna must have had a decisive influence on the diversification of the plants themselves and also on the evolution of phytophagous insects. Coevolutionary mechanisms, such as the production of a new alkaloid provide a means to fight back, in turn causing food selection. Ehrlich and Raven (1964) also proposed that the specific contact between plants and herbivores could be the main area of interaction responsible for the appearance of organic diversity in a community of terrestrial organisms inhabiting a microhabitat.

We know that insects, except for some butterflies, are blind to red. We have cited previously the case in Papua New Guinea of white rhododendrons visited by night moths and red species by day flying birds. These examples can be multiplied. For example, species of *Aquilegia* are interesting because the red species, *A. formosa*, is pollinated by hummingbirds, and the sympatric white species, *A. pubescens*, by night flying hawk moths. The color, flower position, and length of nectariferous spur, vary in function for these pollinators. Hybrids produced by bumble bee cross-pollination probably exist and these also may be pollinated by birds or hawk moths. Variations are still occurring, resulting in secondary hybrids and further discrimination by pollinators. If so, insects can direct the evolution of species, but the contrary is equally true because plants have a real influence on the speciation of pollinators.

Also note that practically all plants, even *Ginkgo biloba*, and species of *Eucalyptus*, have enemies, at least in their native habitat. Only some rarities, chemically well protected, seem to be immune from insect attack, such as *Melia azedarach*, the Persian lilac (Meliaceae), but is this certain? The ginkgo, where it is still growing wild in the Chinese mountains, has its enemies. No plant, however, is attacked by all the phytophagous insects living in its area. If that were so, the plant probably would be condemned to immediate extinction. The neem trees (*Azadirachta indica and Melia azedarach*) endemic in tropical Asia, produce several compounds which seem to be toxic, or at least have an antifeeding effect on most insects. However, some insects that do feed on these trees are: *Melolontha* spp. (Coleoptera), scale insects, mealybugs, bark eating caterpillars (*Indarbela* spp.), and spider mites (Kranz, et al., 1977).

A question that is often raised is: What is the relationship, if any, between primitive insects and plants? In other words, would a primitive insect feed on primitive plants, and would this be indicative of the primitive nature of the plant, or vice versa? It is evident that we are tempted sometimes to answer "yes." Caterpillar species of the genus *Micropteryx,* as previously mentioned, a somewhat aberrant genus between Trichoptera and Lepidoptera, live on hepatica (liverworts). These caterpillars, with very long antennae, are among the most primitive of the Lepidoptera. We have searched for some similar examples among the Chrysomelidae (Coleoptera) which, as the Lepidoptera, must have appeared with the angiosperms during the lower Cretaceous. There are some archaic Sagrinae or related groups in Australia and America which live on cycads, very primitive vascular plants, but in general, the Chrysomelidae on conifers, vascular cryptogams (such as horsetails and ferns), and mosses, are very highly evolved species which have abandoned the angiosperms through secondary adaptations. The same rule is verified by the fern Microlepidoptera (some Pyralidae), which belong to highly evolved species. There are very few insects on ferns, perhaps because of their hormone derivatives (phytohormones), certain repulsive compounds, which many species produce.

We don't know of any butterfly (Papilionoidea) which feeds on bryophytes or vascular cryptogams and only certain moths, such as species of *Papaipema* (Noctuidae) do so. A few butterflies live on gymnosperms including the Cycadaceae, and in this case, there are Lycaenidae which also live on angiosperms. Here also the antiquity of the insect group has nothing to offer in parallel with the antiquity of the plant. Lepidoptera on gymnosperms probably were derived from ancestors living on angiosperms, as well as ones on monocotyledons (except perhaps the Morphidae). These are derived from ancestors living on dicotyledons. It remains to be demonstrated, however, whether the dicotyledons or the monocotyledons are the most primitive group. Both evolved during the Cretaceous.

There is the possibility of coevolution between arthropods and pterydophytes (Gerson, 1979), before and after the appearance of the angiosperms, which seems to be explained by the presence on ferns of both primitive and derived insect species. It may be, however, that ferns are actually under utilized by insects. Most of the ferns attacked are species of Polypodiaceae, which are sometimes associated with ants. Also, it must be noted, the older insect orders are better represented on ferns than the more recent ones.

The true significance of the previous example, of cantharophily (flower attraction for beetles) by the cycads and primitive ly (flower attraction for beetles) by the cycads and primitive angiosperms, is still doubtful. Certainly this may be an old relationship, but still to be proven. Finally, except for the pollinators, a rule, linking the lower insects to the lower plants, and *vice versa*, is not evident. A primitive species may very well live on a derived plant species and the reverse is true. This fact can be easily understood for leaf miners since their adaptation to that diet came very late in their evolution when the flowering plants were already abundant and diversified and their reconquest of the lower plants came about sporadically.

Finally, we must assert that nonphytophagous species can be sometimes associated with plants for some unknown reason. For instance, we have met once in Phu Kae, Saraburi, Thailand, located between Bangkok and the National Park of Khoa Yai, an enormous aggregation (15 million, or 170 kg) of three species of tenebrionid beetles crawling over high trees including *Aegle marmelos, Dipterocarpus elatus, D. indicus, Schleichera oleosa, Stereospermum chelonoides,* and *Sterculia foetida.* Each tree was covered from the trunk to the top by 150 to 170 kg of small insects belonging to the genus *Mesomorphus.* How to explain this observation? Jolivet (1971) stated that there was no relationship with the rainy season. Was it an unknown meteorological phenomenon or a special trophic relationship? Being saprophagous, the beetles were feeding on the bark or lichens. I have seen enormous aggregations of species of *Calosoma* in Senegal, but in this case this was directly linked with the beginning of the rainy season. These examples remain mysteries, and there is yet much to be said about coevolution and its possible ramifications.

"Fust time I see de boll weebil sitting on a square
Nex' time I see de boll weebil he got his family dere."

-Old American folk song.

CHAPTER 15

Conclusions

In the previous chapters we have discussed insect/plant coadaptations. These are certainly mutual adaptations, often complex and indispensable, particularly in the case of orchids, myrmecophilous, and carnivorous plants. We have tried to impartially present this coevolution, fruit of interrelationships and long selection. The great Darwin himself was astonished by the adaptation between an orchid and its pollinator. The more we advance in our research, the more we see, for instance, in the example of species of *Heliconius,* that all these adaptations are extraordinarily complex and that the simple and "primary" schemes of Gaston Bonnier are very much out of date.

The complexity of the species of *Heliconius* is double: food of the caterpillar, *Passiflora* spp., food of the adult (cucurbit flowers). All this complexity has developed harmoniously at the price of a considerable geographical variation between the eater and the eaten. The behavior of *Heliconius* spp. on *Passiflora* spp. in tropical America is not so different from certain species of Malaysian *Ornithoptera* living on species of *Aristolochia*. Certain species are oligophagous (in several species of *Aristolochia* [Dutchman's Pipe]), others strictly monophagous, which explains their scarcity after the clearing of the forest.

Plants are at the base of all plant-animal food chains, but only in some groups does this interdependence become symbiotic or even pseudosymbiotic. There are insects which show us the most extraordinary adaptations of this.

Not all of the so-called phytophagous animals always eat living plants, at least at every stage. Goats and cows, particularly in desert areas of the Sahara or in Arabia, sometimes eat paper. Paper is, of course, made of cellulose. This is chewed mainly as a reflex. Certainly the bandages that a goat near a hospital in Djibouti seemed to eat with delection must have been indigestible. Stored product beetles often eat both plant and animal material.

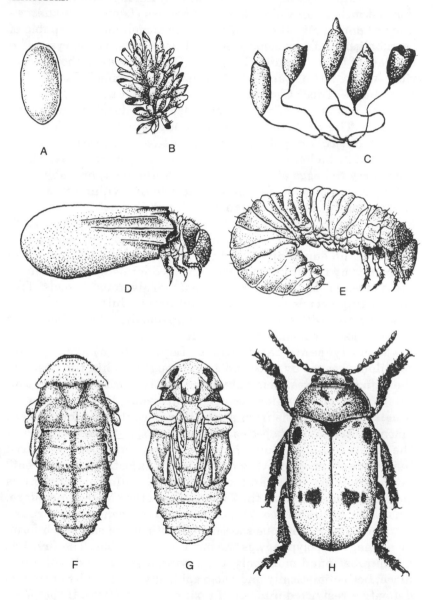

FIG. 27. Some carnivorous "phytophagous" beetles (Chrysomelidae, Clytrinae): eggs, larvae, pupae, and adults of *Clytra quadripunctata*, except C are eggs of *Labidostomis tridentata* (after Jolivet, 1952).

Aberrant cases of feeding are known among various insects. For instance, adult Colorado potato beetles *(Leptinotarsa decem-lineata)* and many other species of Chrysomelidae are capable of cannibalism and occasionally eat their own eggs. The chrysomelid beetles, of the subfamily Clytrinae (fig. 27), are another exception. All the larvae (and some adults) are myrmecophilous, *i.e.*, host of ants. These Clytrinae live in great part at the expense of the eggs and larvae of ants, eating also plant debris found inside the nest. The clytrines with orange and black adults (an aposematic device) are polyphagous on bushes and grasses and lay their eggs, which mimic seeds, in the vicinity of ant hills. These ants carry the eggs of the beetles inside the nest, probably mistaking them for seeds. The hatched larvae live within cases made of excreta, the surface design of which is characteristic of each species. The front part is open to allow the head to emerge first. Often within the nest the larvae pull their heads back inside the case and the empty space at the end of the tube is often filled with eggs by the ant hosts. The eggs are immediately eaten by these strange "phytophagous" turned "carnivorous" insects. This same thing occurs with myrmecophilous adult clytrines and probably also with termitophilous chrysomelids, but these cases are rare, localized, and poorly known.

Many lycaenid caterpillars (Lepidoptera) are myrmecophilous, reared by ants and even brought within the nests by the ants themselves. Being phytophagous on dicotyledons, certain caterpillars of lycaenids become carnivorous, lichenivorous, or fungivorous, and this from a totally different selection process among closely related species. It is evident that the caterpillars of the moths living on sloths have a food very different from free living related species living on the host's excreta. The adult eats desquamations of the animal's skin. We could find numerous similar examples among the Tineidae and other Microlepidoptera. One Latin American bee, *Trigona hypogea,* is even necrophagous.

Among phytophagous species, it is difficult to say which form, monophagy or polyphagy, is the most advantageous. The first has an unquestioned monopoly, occupying a restricted ecological niche, both climatically and geographically. It saves its energy to detoxify a restricted number of toxins (Feeny, 1975). If the plant is dying, the specialist dies with it, since there is little chance that it could eat something else. The polyphagous, or oligophagous, species, *i.e.*, the generalists, have a much greater margin of choice. These also have a wider distribution, hence more chance of survival, and greater plasticity at the cost of a supplementary metabolic expense, since the toxins to be neutralized vary from one plant to another.

The true phytophagous species represent about 50% of the living insects, and evidently during their evolution, phytophagy must have appeared first, followed by the carnivores which fed on them. We can see from previously cited examples how different are the associations between insects and plants. However, as was said in the introduction, many problems have not been treated: the role of plant pests in crop destruction; transmission of cryptogamic, mycoplasmic, bacterial, rickettsial, or viral diseases by insects; useful insects such as bees, cocheneal, and silkworms which produce from plants useful substances; toxic plants which transmit their toxicity to insects and protect them indirectly (as in zygaenids, chrysomelids, and danaids); problems of homochromy, or protective resemblance between an insect and a part of a plant; the symbiont problem necessary to certain phytophagous species (mostly those that are xylophagous) and for hematophagous species; vitamins, hormones, and other topics.

However, it must be stated that toxicity of plants is not always responsible for toxicity in insects. Species of *Timarcha*, big black chrysomelids, called bloodynose beetles, may feed on Rubiaciaceae or plantains. In either case they produce the same toxic compounds during the autohemorrhage. Pasteels and Deloze (1977) have recently demonstrated that cardiac glucosids effect protection against predators. These are not necessarily taken in by chrysomelid beetles from their host plant, but can be entirely synthesized if the host plant does not contain it. Species of Danaidae, on the contrary, obtain from their host plants (Asclepiadaceae, Apocynaceae, or Moraceae) not only the cardenolids, toxic to predators, but also the pheromone precursors.

A species of *Danaus*, fed on cabbage, loses its emetic agents and becomes edible, something that birds would learn rapidly if this happened in nature. They have recently learned that only the outer tissue of the larvae are poisonous, and that they can, in the case of blue jays, scoop out and eat the inner parts of the larvae with safety. A quantity of mimics imitate species of danaids and heliconids, or both imitate each other (Müllerian and Batesian mimicry). It is the phenomenon of a "sheep in wolf's clothing" for the nontoxic species, while in the case of aggressive mimicry, it is the wolf that carries the sheep's skin to get its prey. Grasshoppers taken from species of *Calotropis* (Asclepiadaceae), fed on lettuce or carrots, lose practically all of their toxic cardenolids, while still aposematic and often capable of the emission of blood (hemaphrorrhea), and air to repel the eventual predator, generally a bird. We see that in this field we must not generalize.

In aggressive (Peckhamian) mimicry, a phytophagous insect (in this example, a species of Chrysomelidae, Alticinae) is imitated by a carnivorous carabid (Lebiina) beetle which attacks and feeds on the alticine larva or adult. The "wolf" then carries the "sheep's clothing" to reach and capture its prey and thereby gets poisoned by the prey. So toxic are the pupae of these alticine victims and the lebinne predators that both are used as arrow poison by Kalahari Kung. Examples of Batesian and Müllerian mimicry are common among Coleoptera in the tropics where, generally, phytophagous species are involved, because they are toxic and aposomatic, and their toxicity is often due to their food plant.

The aposematic insects, *i.e.*, those with a color indicating their toxicity (also called vexillar), such as the danaid butterflies, can bring some benefits to their host plant, such as a warning to eventual herbivores, or by reinforcing the protective odor of the plant. These advantages seem, however, small if we compare them with the ones the insects get from its toxic host plant. The quality of the food of silkworms also influences indirectly the quality and the secretion of their silk. This is so true that, during experiments in Japan of feeding the caterpillars of *Bombyx mori* on synthetic media instead of mulberry, the insect accepted the new food only if that medium contained at least 10% mulberry leaves. In Japan they tried the culturing of mulberry tissue and the results have been positive, provided the culture is produced in light and if the cells contain at least a minimum of 1% of the normal chlorophyll content of the leaves. This experiment shows us the complexity of the acceptance or refusal of specific plants by insects. Synthetic media are also a way to analyze their preferences, the dosage of the attractants, and the need and amount of various chemicals, in rearing insects. In the previous examples, tissue culture with 1% chlorophyll has produced an excellent quality silk. Many other caterpillars are less difficult to feed and will accept a 100% synthetic medium.

Outside of visual, olfactive, gustative, and chemical reasons for host plant selection, the quality of the plant influences also the rapidity of the development and vigor of the insect. Choices of this nature are not so hidden as we might otherwise suppose.

All of this integration of plants and insects into narrow adaptations, such as myrmecophily, carnivorous function, epizoic symbiosis, pollination, and its caprices, must have occurred during a relatively short evolutionary period, if we propose that the majority of these phenomena deal with the angiosperms and that these plants appeared only at the end of the Jurassic and the beginning of the Cretaceous. It is not even too early for the orchids. By the Cenozoic, all the present day associations already existed and have not been greatly modified since then.

It is evident that food selection has a strong influence on the species genesis, *i.e.*, speciation. Currently there exists in nature a whole series of biological races and developing species, not only among leaf miners, but also among externally phytophagous species, as well as many other species. Selection is often the function of existing preadaptations to a given diet. Polyphagous species have more chances to evolve quickly than monophagous species.

Roughly, plant diversity, as outlined by Hutchinson (1964), leads to increased diversity of phytophagous animals, but the contrary is equally true. The diversity of tropical species is due to climatic conditions, permanent and optimal, which allowed development of the great diversity of phytophagous insects as well as the plants on which they feed. There should not be any confusion between the superficial and reversible variations, *i.e.*, somatic variations in relation to the biochemical and seasonal changes of the host plant (cassids, nymphalid butterfly species of *Vanessa)* and with the actual biological races which are ecologically and morphologically stable and genetically transmitted.

Finally, a great many insects are detriticolous or herbivorous. Among the herbivores, some became selective. It is that passage from polyphagy to oligophagy, then to monophagy, which has produced, by a series of restrictive mutations, the ecological potentialities of the insects (fig. 28). Some (*e.g.*, leaf miners) can sometimes partially recuperate their initial potentials. Anyhow, the most striking adaptation, fruit selection, is instinctive behavior which delimits the choice of the host plant.

It is this mutual evolution (coevolution) during past geological ages of insects and plants, their interactions, their reciprocal adaptations, symbioses, defenses, nutrition, and protection, which we have tried to review in this book. The actions and reactions are so varied, so complex, and the subject so vast that the previous chapters have only been an introductory summary to what surely needs to be treated in as many separate books.

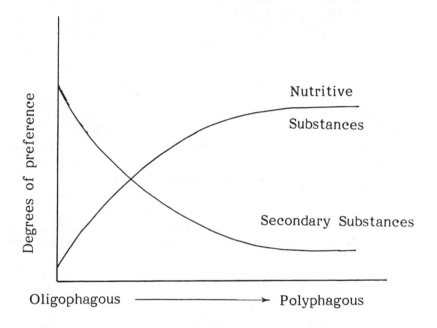

FIG. 28. Hypothetical relationships between the degree of preference of the phytophagous insects and the basic plant substances and their degree of polyphagy (after Edwards and Wrattan, 1981).

BIBLIOGRAPHY

Note: The first series of references are those of general interest and supplementary reading. The second series are articles and specific references mentioned in the text.

A. General

Ahmad, S., *et al.* 1983. *Herbivorous insects, host seeking behavior and mechanism.* New York: Academic Press, 257 pp.

Ananthakrishnan, T. N. 1984. *Biology of gall insects.* London: Edward Arnold, 362 pp.

Atkins, M. D. 1980. *Introduction to Insect Behavior.* New York: Macmillan Publ. Co., 237 pp.

Balachowsky, A. S. 1951. *La lutte contre les Insects.* Paris: Ed. Payot.

Blum, M. S. 1981. *Chemical defenses of Arthropods.*New York: Academic Press, 562 pp.

Brues, C. T. 1946. *Insect dietary.* Cambridge: Harvard Univ. Press., 466 pp.

Chapman, R. F. and Bernays, E. A. 1978. *Proceeding of the 4th International Symposium on Insects and Host Plants.* Slough, England, pp. 201-766.

Chrispeels, M. and Sadavu, D. 1977. *Plants, Food and People.* San Francisco: Freeman, 278 pp.

Conn, E. E. (Ed.), 1981. *Secondary Plant Products in the Biochemistry of Plants*, vol. 7. New York: Academic Press, 798 pp.

Cooper, E. 1965. *Insects and Plants.* London: Butterworth Press, 142 pp.

Darwin, C. 1875. *Insectivorous Plants.* New York: Appleton and Co., 462 pp.

Edwards, P. J. and Wratten, D. S., 1980. *Ecology of Insect-Plant Interactions.* London: Arnold, 60 pp.

Elton, C. S. 1958. *The Ecology of Invasions by Animals and Plants.* London: Methuen & Co., 181 pp.

Faegri, K. and Van der Pijl, L., 1966. *The Principles of Pollination Ecology.* New York: Pergamon Press, 291 pp.

Forey, P. L. 1980. *The Evolving Biosphere.* London: British Museum (Natural History), 305 pp.

Free, J. B. 1970. *Insect Pollination of Crops.* New York: Academic Press, 544 pp.

Futuyma, D. J. and Slatkin, M. (Eds.). 1983. *Coevolution.* Sunderland, MA: Sinauer Assoc., x+555 pp.

Gilbert, L. E. and Raven, P. M. 1975. *Coevolution of Animals and Plants.* Austin: Univ. Texas Press, 246 pp.

Grassé, P. P. 1977. *Cecidies, in* Traité de Zoologie. Paris: Masson, vol. 8, fasc. 5, pt. 2, 496 pp.

Grassé, P. P. 1976. *Insectes et Fleurs, in* Traité de Zoologie. Paris: Masson, vol. 8, fasc. 4, 984 pp.

Hardone, J. B. 1972. *Phytochemical Ecology.* New York: Academic Press, 272 pp.

Hardone, J. B. 1978. *Biochemical Aspects of Plant and Animal Coevolution.* London: Academic Press, 425 pp.

Hartzell, A. *In* S. Mark Henry 1967. *Insect Ectosymbiosis* in *Symbiosis.* New York: Academic Press, pp. 107-140.

Heinrich, B. 1979. *Bumble Bee Economics.* Cambridge: Harvard Univ. Press, 245 pp.

Hering, M. 1951. *Biology of the Leaf Miners.* The Hague: Junk, Publ., 420 pp.

Heywood, V. M. 1975. *Taxonomy and Biology.* New York: Academic Press, 370 pp.

Houard, C. 1908-1913. *Les Zoocecidies des plantes d'Europe et du Bassin de la Mediterranee.* Paris: Herman ed.

Hutchinson, J. 1964. *Evolution and Phylogeny of Flowering Plants.* New York: Academic Press.

Huxley, A. 1974. *Plant and Planet.* London: Allen Lane, 428 pp.

Janzen, D. M. 1975. *Biology of Plants in the Tropics.* London: Arnold Ed., 66 pp.

Jennings, D. M. 1975. *Symbiosis.* Cambridge: Cambridge Univ. Press, 633 pp.

Jermy, J. *et al.* 1976. *The Hostplant in Relation to Insect Behavior and Reproduction.* New York: Plenum Press, 322 pp.

Jolivet, P. 1980. *Les Insectes et l'Homme.* Paris: Presses Univ. Fr., 129 pp.

Kasasian, L. 1971. *Weed Control in the Tropics.* London: Leonard Hill, 307 pp.

Kullenberg, B. 1961. *Studies in Ophrys Pollination.* Uppsala: Almqvist & Wiksells, 34 pp.

Labeyrie, V. 1977. *Camportement des Insectes en lieu Trophique.* Paris: CNRS, 493 pp.

Labeyrie, V. 1981. *The Ecology of Bruchids attacking Legumes (Pulses).* The Hague: Junk Ed., 233 pp.

Larson, P. P. and M. W. 1965. *Ants Observed*. London: Scientific Book Club, 192 pp.

Lloyd, F. E. 1942. *The Carnivorous Plants*. New York: Ronald Press Co., 352 pp., pls.

Louveaux, J. 1965. *Plantes Carnivores et Vegetaux Hostiles*. Paris: Hachette, 108 pp.

Mani, N. S. 1964. *The Ecology of Plant Galls*. The Hague: Junk, Ed., 434 pp.

Mark Henry, S. 1966-1967. *Symbiosis*. New York: Academic Press, 2 vols., 478 pp.; 443 pp.

Meeuse, B. J. D. 1961. *The Story of Pollination*. New York: Ronald Press, 243 pp.

Metcalf, R. L., and Luckmann, W. H. 1975. *Introduction to Insect Pest Management*. New York: Wiley & Sons, 587 pp.

Mound, L. A., and Waloff, N. 1978. *Diversity of Insect Faunas*. London: Roy. Ent. Soc., 204 pp.

National Academy of Sciences. 1969. *Insect Plant Interactions.*, Washington, DC: Nat. Acad. Sci., 98 pp.

Painter, R. M. 1951. *Insect Resistance in Crop Plants*. New York: Macmillan, 520 pp.

Price, P. W. 1985. *Insect Ecology*, 2nd. ed. New York: Wiley & Sons, 607 pp.

Proctor, M. and Yeo, P. 1979. *The Pollination of Flowers*. London: Collins, 418 pp.

Richard, A. J. R. 1978. *The Pollination of Flowers by Insects*. London: Academic Press, 213 pp.

Ridley, H. N. 1930. *The Dispersal of Plants throughout the World*. Reeve.

Rodriguez, J. G. 1972. *Insect and Mite Nutrition*. Amsterdam: North Holland, 702 pp.

Schnell, R. 1970. *Introduction a la phytogeographie des pays tropicaux*. Paris: Gauther Villars Ed., 500 pp.

Sondheimer, E. and Simeone, J. B. 1970. *Chemical Ecology*. New York: Academic Press, 336 pp.

Schremner, F. 1976. *Other Interactions of Animals and Plants*. In: Grzimek's Encyclopedia of Ecology. New York: Van Nostrand Reinhold, 705 pp.

Souchon, C.1965. *Les Insectes et les Plantes*. Paris: Presses Univ. de France, Que saisje?, 123 pp.

Strong, D. R., *et al*. 1984. *Insects on Plants: Community Patterns and Mechanisms*. Cambridge: Harvard Univ.Press, 313 pp.

Swanton, E. W. 1912. *British Galls*. London: Methuen & Co., 287 pp.

Torma, M., *et al.* 1958. *First Symposium on Insect and Food Plants*. Amsterdam: Ent. Exp. & Appl. I (1 & 2), pp. 1-152.

Van der Pijl, L. 1982. *Principles of Dispersal of Higher Plants*. New York: Springer Verlag, 215 pp.

Van der Pijl, L. and Dodson, C. M. 1966. *Orchid Flowers: Their Pollination and Evolution*. Coral Gables: Univ. Miami Press, 214 pp.

Van Emden, H. P. 1973. *Insect Plant Relationships*. London: Symp. Roy. Ent. Soc. London, vol. 6, 215 pp.

Wallace, J. M. and Mansell, R. L. 1976. *Biochemical Interaction between Plants and Insects*. New York: Plenum Press, 425 pp.

Wheeler, W. M. 1925. *Ants. Their Structure, Development and Behavior*. New York: Columbia Univ. Press, 633 pp.

B. Speciality References

Aoxi, J. I. 1966. *Epizoic Symbiosis: An Oribatid Mite, Symbioribates papuensis, representing a new family, from cryptogamic plants growing on backs of Papuan weevils*. Pacific Insects, 8 (1): 281-289.

Arditti, J. 1966. *Orchids*. Scientific American, 214 (1), pp. 70-78.

Baker, H. G. and Baker, I. 1973. *Amino-acids in nectar and their evolutionary significance*. Nature, 24:543-545.

Batra, S. W. T. and L. R. 1967. *Fungus Gardens of Insects*. Scientific American, 217(5): 112-120.

Batra, L. R. 1979. *Insect Fungus Sgmbiosis in Mutualism and Symbiosis*. New York: Wiley & Sons, 276 pp.

Beck, S. D. 1965. *Resistance of plants to insects*. Ann. Rev. Ent., 1: 207-232.

Bentley, B. L. and Elias, T. 1983. *The Biology of Nectaries*. New York: Columbia Univ. Press, 259 pp.

Benzing, D. H. 1970. *An investigation of two Bromeliad Myrmecophytes, Tillandsia butzii Mez, T. caputmedusae E. Morron and their ants*. Bull. Torrey Bot. Club, 97 (2): 109-115.

Bequaert, J. 1921. *Ants in their diverse relations to the plant world. In* Wheeler, *The Ants*, Bull. Ann. Mus. Nat. Hist., 45: 333-348.

Blatter, E. 1928. *Myrmecosymbiosis in the Indo-malayan Flora*. J. India Bot. Soc., 7: 176-185.

Boer, G. de, *et al.* 1977. *Chemoreceptors in the preoral cavity of the tobacco hornworm, Manduca sexta, and their possible function in feeding behavior.* Ent. exp. & appl., 21: 287-298.

Brues, C. T. 1924. *The specificity of foodplants in the evolution of phytophagous insects.* Amer. Nat., 58: 127-144.

Brues, C. T. 1951. *Insects in amber.* Scientific American, 6 pp.

Clute, W.N.1925. *Maneating trees.* American Botanist, 31: 70-73.

Crawley, M. J. 1983. *The Herbivory: The Dynamics of animal-plant Interactions.* Berkeley: Univ. Calif. Press, 437 pp.

Denno, R. F. and McClure, M. S. 1983. *Variable plants and herbivores in natural and managed systems.* New York: Academic Press, 717 pp.

Dethier, V. G 1954. *Evolution of Feeding preferences in phytophagous insects.* Evolution, 8(1): 33-54.

Dethier, V. G. 1960. *The designation of chemicals in terms of responses they elicit from insects.* J. Econ. Ent., 53: 134-136.

Dethier, V. G. *In* Sendheimer and Simeone. 1970. *Chemical interactions between plants and insects.* Chemical Ecology. New York: Academic Press, pp. 83-102.

Dethier, V. G. 1977. *The role of chemosensory patterns in the discrimination of food plants.* Colloques Int. CNRS, Paris. 265: 103-114.

Dethier, V. G.1977. *Gustatory sense of complex mixed stimuli by insects.* Olfaction & Taste, Paris. 323-331.

Dethier, V. G. 1978. *Other tastes, other worlds.* Science, 201: 224.

Dethier, V. G. 1978. *Studies on insect host plant relations - past and future.* Ent. exp. & appl., 24: 559-566.

Dethier, V. G. and Yost, M. T. 1979. *Oligophagy and absence of food-aversion learning in tobacco hornworms, Manduca sexta.* Physiol. Ent. 4: 125-130.

Duranton, P., *et al.* 1982. *Manuel de prospection acridiennone en zone tropicale.* Paris: Gerdat, Seche, vol. 2, 695 pp.

Eastop, V. F. 1981. *Coevolution of plants and insects. In* Forey, P. L., *The evolving biosphere.* Cambridge: Cambridge Univ. Press, 179-190.

Ehrlich, P. R. and Raven, P. H. 1965. *Butterflies and Plants: a study of coevolution.* Evolution, 18: 586-608.

Feeny, P. 1976. *Plant apparency and chemical defense. In* Wallace and Mansell. *Biochemical interelation between plants and insects.* New York: Plenum Press, pp. 1-40.

Feeny, P. 1977. *Defensive ecology of the Cruciferae.* Ann. Missouri Bot. Garden, 64: 221-234.

Fraenkel, G. 1959. *The raison d'etre of secondary substances.* Science, 129: 1466-1470.

Fraenkel, G. 1969. *Evolution of our thoughts on secondary plant substances.* Ent. exp. & appl., 12: 473-486.

Gallun, R. L. 1972. *Genetic relationships between host plants and insects.* J. Environ. Qual., 1: 259-265.

Gerson, U. 1969. *Moss-Arthropod associations.* Bryologist, 72: 495-500.

Gerson, U. 1974-1976. *The associatians of Algae and Arthropods.* Rev. Algologique, 2: 18-41; 213-247.

Gerson, U. 1973. *The associations between pteridophytes and arthrapods.* Fern.Gaz., 12(1): 29-45.

Gressitt, J. L. *et al.,* 1965. *Flora and fauna on backs of large Papuan moss-forest weevils.* Science, 150 (3705): 1833-1835.

Gressitt, J. L. 1966. *The weevil genus Pentorhytes (Col.) involving cacao pests and epizoic symbiosis with cryptogamic plants and microfauna.* Pacific Ins. 8(4): 15-65.

Gressitt, J. L. 1966. *Epizoic Symbiosis: The Papuan weevil genus Gymnopholus (Leptopiinae) symbiotic with cryptogamic plants, oribatid mites, rotifers and nematodes.* Pacific Ins., 8(1): 221-280.

Gressitt, J. L. 1966. *Epizoic Symbiosis: Cryptogamic plants growing an variaus ueevils and an a oolydiid beetle in New Guinea.* Pacific Ins., 8(1): 294-297.

Gressitt, J. L. 1969. *Epizoic Symbiosis.* Ent. News, 80(1): 15.

Gressitt, J. L. 1970. *Papuan weevil genus Gymnopholus: second supplement with studies in epizoic symbiosis.* Pacific Ins., 12(4): 753-762.

Gressitt, J. L. 1977. *Papuan weevil genus Gymnopholus: third supplement with studies in epizoic symiosis.* Pacific Ins., 17(23): 179-185.

Gressitt, J. L. and Sedlacek, J. 1967. *Papuan weevil genus Gynnopholous: Supplement and further studies in epizoic symbiosis.* Pacific Ins., 3: 481500.

Heslop-Harrison, Y. 1976. *Carnivoraus plants a century after Darwin.* Endeavour, 35(126): 114-122 .

Heslop-Harrison, Y. 1978. *Carnivorous plants.* Scientific American, 238(2): 104-115.

Hocking, B. 1970. *Insect associations with swollen thorn acacias.* Trans. Roy. Ent. Soc., London, 122(7): 211-255.

Hovanitz, W. 1959. *Insects and plant galls.* Scientific American, Nov., pp. 151-162.

Hsiao, T. H. 1978. *Host plant adaptations among geographic populations of the Colorado Potato Beetle.* Ent. exp. & appl., 24: 437-447.

Jaeger, P. 1971. *Contribution a l'etude de la biologie florale des Asclepiadacees, le Calotropis procera.* Bull. IFAN Dakar, 33(Al): 32-43.

Janzen, D. H. 1966. *Coevolution of mutualism between ants and acacias in Central America.* Evolution, 20(3): 249-275.

Janzen, D. H. 1967. *Interaction of the Bull's Horn Acacia (Acacia cornigera) with ant inhabitant (Pseudomyrmex ferruginea) in Eastern Mexico.* Kans. Univ. Sci. Bull., 47: 315-558.

Janzen, D. H. 1969. *Allelopathy by Myrmecophytes: the Ant Azteca as an allelopathic agent of Cecropia.* Ecology, 50(1): 147-153.

Janzen, D. H. 1972. *Protection of Barteria (Passifloraceae) by Pachysima ants (Pseudomyrcecinae) in a Nigerian rain forest.* Ecology, 53(5): 885-892.

Janzen, D. H. 1974. *Epiphytic myrmecophytes in Sarawak: Mutualism through the feeding of plants by ants.* Biotropica, 6(4): 237-259.

Jolivet, P. 1954. *Phytophagie et selection trophique.* Livre. Jub. Van Straelon, 2: 1101-1134.

Jolivet, P. 1971. *La Nouvelle Guinee Australienne: Introduction ecologique et entomologique.* Paris: Gah. Pacific, 15: 41-70.

Jolivet, P. 1973. *Les Plantes Myrmecophiles du Sud Bst Asiatique.* Gah. Pacific, 17: 41-65.

Jolivet, P. 1979. *Les Chrysomelidae (Coleoptera) des Citrus et apparentes (Rutaoeae) en zone temperee et tropicale.* Bull. Mens. Soc. Linn. Lyon. 43(4): 197-200; 249-256.

Jolivet, P. 1980. *Les mannes: Entomologie et Botanique.* Bull. Soc. Linn. Lyon, Suppl., 49(9) 17-22.

Kennedy, J. S. 1953. *Host plant selection in Aphididae.* Trans. IXth Int. Congress Ent., 2: 106-110.

Kennedy, J. S. and Stroyna, H. L. G. 1959. *Biology of Aphids.* Ann. Rev. Ent., 4: 139-160.

Kennedy, J. S. 1965. *Mechanisms of host plant selection.* Ann. Appl. Biol., 56: 317-322.

Kevan, P. G., Chaloner, W. G., and Savile, D. B. 0. 1975. *Interrelationships of early Terrestrial Arthropods and Plants.* Palaeontology, 18(2): 391-417.

Kogan, M. 1976. *The role of chemical factors in insect/plant relationships.* Proc. 15th Int. Congr. Ent., Washington: 211-227.

Krantz, J. *et al.* 1977. *Diseases, pests, and weeds in tropical crops.* Berlin: Paul Parley, Publ., 666 pp.

Manning, A. 1956. *The effect of honey guides.* Behavior, 9: 114-139.

Merrill, E. D.1981. *Plant life in the Pacific World.* Boston: Tuttle Co., 297 pp.

Meeuse, B. J. D. 1969. *Comment les fleurs guident les insectes.* Atomes, 266: 351-357.

Moldenke, A. R. 1979. *Hostplant coevolution and the diversity of bees in relatian to the flora of North America.* Phytologia, 43(4): 357-419.

Monod, T. and Schmitt, C. 1968. *Contribution a l'etude des pseudo-galles formicaires chez quelques acacias africains.* Bull. Ifan, 30(A3): 953-1012, pls.

Mesquin, T. 1971. *Competition for pollinators as a stimulus for the evolution of the flowering time.* Oikos, 22: 398-402.

Nielsen, J. K 1978. *Host plant discrimination within Cruciferae: Feeding responses of four leaf beetles (Col. Chrys.) to glucosinolates, cucurbitacins and cardenolides.* Ent. exp. & appl., 24(1): 41-54.

Painter, R. H. 1958. *Resistance of plants to insects.* Ann. Rev. Ent., 3: 267-290.

Pomomarenko, A. G. 1976. *A new insect from the Cretaceous of Transbaikalia, a possible parasite of pterosaurians.* Paleont. Zhur., 3: 102-106.

Rees, C. J.C. 1969. *Chemoreceptor specificity associated with choice of feeding site by the beetle Chrysolina brunsvicensis on its food plant, Hypericum hirsutum.* Ent. exp. & appl., 12: 565-583.

Rehr, S. S. *et al.* 1973. *Chemical defense in Central American nonant acacias.* J. Anim. Ecol., 42(2): 405-416.

Rohdendorf, B. B. and Raznitsin, A. P. 1980. *The Historical Development of the class Insecta.* (In Russian.) Trudy Paleont. Inst. (Moscow), 175: 1-268.

Samuelson, G. A. 1966. *Epizoic symbiosis: a new Papuan colydiid beetle with epicuticular growth of cryptogamic plants (Col. Colydiidae).* Pacific Ins., 8(1): 250-293.

Saxena, K. N. and Goyal, S. *Host-plant relations of the citrus butterfly Paplio demeleus L: orientation and ovipositional reponses.* Ent. exp. & appl., 24(1): 1-10.

Schnell, R. 1966. *Remarques morphologiques sur les "Myrmecophytes."* Bull. Soc. Bot. France, Memoires, 121-132.

Schnell, R. and de Beaufort, G. 1966. *Contribution a l'etude des plantes a myrmecodomaties de l'Afrique Intertropicale.* Mem. Ifan, 75: 1-66.

Schoonhoven, L. M. 1968. *Chemosensory bases of host plant selection.* Ann. Rev. Ent., 13: 115-136.

Schoonhoven, L. M. 1969. *Gustation and food plant selection in some lepidopterous larvae.* Ent. expl. & appl., 12: 55-56.

Schoonhoven, L. M. 1974. *Studies on the shootborer Hypsila grandella (Zeller). 23. Electroantennogams (EAG) as a tool in the analysis of insect attractants.* Turriolba, 24(1): 24-28.

Slansky, F. 1972. *Latitudinal gradients in species diversity of the New World swallowtail butterflies.* J. Res. Lepid., 2(4): 201-217.

Slansky, F. 1978. *Phagism relationships among butterflies.* J. New York Ent. Soc., 15 pp.

Slocum, R. D. and Lawrey, J. D. 1976. *Viability of the epizoic lichen flora carried and dispersed by green lacewing (Nodita pavida) larvae.* Canadian J. Botany, 54(15): 1827-1831.

Smiley, J. 1978. *Plant chemistry and evolution of host specificity: New evidence from Heliconius and Passiflora.* Science, 201: 745-747.

Teuscher, H. 1956. *Myrmecodia and Hydnophytum.* Nat. Hort. Mag., 35: 49-51.

Teuscher, H. 1967. *Dischidia pectenoides.* Amer. Hort. Mag., 46: 36-40.

Thorsteinson, A. J. 1953. *The chemical sense in phytophagous insects.* Redia, 38: 369-374.

Thorsteinson, A. J. 1953. *The role of host selection in the ecology of phytophagous insects.* Canadian Ent., 85: 276-282.

Thorsteinson, A. J. 1960. *Host plant selection by phytophagous insects.* Ann. Rev. Ent., 5: 193-218.

Van Emden, H. P. 1960. *Plant insect relationships and pest control.* World Rev. Pest Control, 5: 115-123.

Whittaker, R. H. and Feeny, P. P. 1971. *Allelochemics: chemical interactions between species.* Science, 171: 757-770.

Wiens, D. and Prourke, J. 1978. *Rodent pollination in Southern African Protea spp.* Nature, 276: 71-73.

Wilde, J. de and Schoonhoven, L. M., 1969. *Insect and host plant.* Amsterdam: North Holland Pub., 810 pp.

Wilde, J. de. 1983. *Proceedings of the 5th International Symposium on Insect-plant relationships.* Wageningen: Pudoc., 464 pp.

Williams, K. S. and Gilbert, L. E. 1981. *Insects as selective agents on plant vegetative morphology: egg mimicry reduces egg laying by butterflies.* Science, 212: 467-469.

Wilson, C. L. and Graham, C. L. 1983. *Exotic plant pests and North American Agriculture.* New York: Academic Press, 528 pp.

Wisser, J. 1983. *South African parasitic plants.* Johannesburg.

Zwolmer, H. and Harris, G. 1971. *Host specificity determination of insects for biological control of weeds.* Ann. Rev. Ent., 16: 157-178.

INDEX

electroantennogram, 67
Elysia atroviridis, 98
Emex australis, 45
Empoasca fabae, 51
endogenous, 124
endoparasitic, 5, 11
endozoochory, 152
entomocecidia 103
entomochory, 152
entomogamy, 122
Ephedra, 131
Epilobium, 20, 37
epizoochory, 153
Eragrostis teff, 2
Erigeron, 46
ergot, rye 153
ergotism, 153
Eucalyptus, 13, 26, 34, 114, 160
Euchelia jacobaeae, 65
Euglena, 10
  jambos, 46
Eupatorium, 26
  adenophorum, 45
euphagy, 20
Euphorbia, 22, 28, 132
  cyparissias, 51
  hirta, 45
euryphagy, 22
Euspelapteryx phasianipennella, 35
Evodia, 19
  accedens, 20

feeding sites, 51
Fenella voigti, 22
Ficus, 126
  carica, 128
filter feeders, 12
finches, Galapagos, 1
fleas, 11, 156
flies, 49, 60, 77
flora, Gondwanian, 157
floricolous animals, 13
flower, hermaphroditic, 127
  odor, 133
fly, cecidomyid, 105
food chains, 7, 8
  selection, 2
  formula, 56
  origin, 156
  specialization 2
Formica rufa, 154
Fragaria, 24
Fraxinus excelsior, 33
Freycinetia, 122
frondicolous, 13
frugivorous animals 14, 17

Fungi Imperfecti, 112
fungivores, 12, 17
Fuschia, 46

Galago, 122
Galeopsis, 26
Galerucinae, 130
Galinsoga, 44
Galium, 22, 24, 35, 50, 57, 60
  aparine, 50
gall, 103, 114
  apple, 104
  insect groups, 104
  pineapple, 107
gallicolous, 111
  animals, 14
garden, Atta, 114
  cryptogamous, 97
  fungus, 109
gastropod, 98
Gastrophysa, 49
  atrocyanea, 32
  viridula, 57
Gentisea, 75
Geranium, 22
Ginkgo, 152
Ginkgo biloba, 40, 166
Giraudiella, 107
Gloriosa superba, 135
Glossopteris, 156
Gongora, 134, 140, 141
Gongorinae, 134
Goniothalamus riddleyi, 95
Gordioidea, 11
Goryles, 144
Gracilaria syringella, 32
grasshopper, 98, 166
Gryllotalpa, 14
Gurania, 55
Gymnopholus, 98, 102, 101, 103
Gymnosporangium sabinae, 35
gynoecium, 121

Habenaria, 136, 156
Halictus, 149
hausteria, 111
heaven, bread of, 119
hecogamy, 130
Heliamphora, 73
Helianthemum, 30
heliconia, leaf bracts, 71
Heliconius, 55, 148, 163
Heliothis, 57, 66
  armigera, 56
Heliotropium, 43
hematophages, 11